# SpringerBriefs in Environmental Science

For further volumes:
http://www.springer.com/series/8868

Paul A. Montagna · Terence A. Palmer
Jennifer Beseres Pollack

# Hydrological Changes
# and Estuarine Dynamics

 Springer

Paul A. Montagna
Harte Research Institute
Texas A&M University-Corpus Christi
Corpus Christi
TX, USA

Jennifer Beseres Pollack
Life Sciences Department
Texas A&M University-Corpus Christi
Corpus Christi
TX, USA

Terence A. Palmer
Harte Research Institute
Texas A&M University-Corpus Christi
Corpus Christi
TX, USA

ISSN 2191-5547          ISSN 2191-5555   (electronic)
ISBN 978-1-4614-5832-6          ISBN 978-1-4614-5833-3   (eBook)
DOI 10.1007/978-1-4614-5833-3
Springer New York Heidelberg Dordrecht London

Library of Congress Control Number: 2012950862

© The Author(s) 2013
This work is subject to copyright. All rights are reserved by the Publisher, whether the whole or part of the material is concerned, specifically the rights of translation, reprinting, reuse of illustrations, recitation, broadcasting, reproduction on microfilms or in any other physical way, and transmission or information storage and retrieval, electronic adaptation, computer software, or by similar or dissimilar methodology now known or hereafter developed. Exempted from this legal reservation are brief excerpts in connection with reviews or scholarly analysis or material supplied specifically for the purpose of being entered and executed on a computer system, for exclusive use by the purchaser of the work. Duplication of this publication or parts thereof is permitted only under the provisions of the Copyright Law of the Publisher's location, in its current version, and permission for use must always be obtained from Springer. Permissions for use may be obtained through RightsLink at the Copyright Clearance Center. Violations are liable to prosecution under the respective Copyright Law.
The use of general descriptive names, registered names, trademarks, service marks, etc. in this publication does not imply, even in the absence of a specific statement, that such names are exempt from the relevant protective laws and regulations and therefore free for general use.
While the advice and information in this book are believed to be true and accurate at the date of publication, neither the authors nor the editors nor the publisher can accept any legal responsibility for any errors or omissions that may be made. The publisher makes no warranty, express or implied, with respect to the material contained herein.

Printed on acid-free paper

Springer is part of Springer Science+Business Media (www.springer.com)

# Abstract

Nothing is more fundamental to an estuary than the mixing of fresh and salt water, but there has been explosive growth in water diversions for human needs since the middle of the twentieth century. Today more than 60 % of all runoff on Earth is captured, thus water development projects where dams and diversions have been built have altered the environmental flow landscapes. In fact, there is currently more water in reservoirs than natural lakes and rivers.

Estuaries are the most productive environments on Earth, and this is in part due to freshwater inflow from rivers to the coast. The freshwater dilutes marine water and transports nutrients and sediments to the coast. Estuaries are characterized by salinity and nutrient gradients, which are important in regulating nearly all ecological and biological processes in estuaries. Sediment is important for building habitats and balancing erosive forces. However, hydrological patterns are being altered as flow is diverted for human use, and this is reducing flows to the coast in all parts of the developed world. A review of available research shows that this reduction of freshwater flow leads to increased salinities, but in contrast reduced nutrient and sediment loading; and that these alterations are having profound effects on coastal resources, particularly for estuarine dependent species. However, there are still many research gaps, and in particular estuaries represent a continuum, so it is very difficult to identify freshwater needs to maintain ecosystem services.

While many countries have water quality programs, few have policies or regulations that protect water quantity alterations. So, while pollution has generally decreased since the first environmental laws of the 1970s, habitats have been degraded in part because of alterations in hydrological regimes. Where there are laws or rules to protect riverine or estuarine habitats, it is very difficult to identify how much instream flow is needed for a river to function like a river, or how much inflow is needed to maintain the characteristics of the estuary that make up the estuarine signature unique to each estuary. Inflow needs studies generally follow a sequence of steps. The first step is to define the natural and historical flow regimes and how the rivers have contributed to salinity zonation in the estuaries. The second step is to define the marine biological resources to be protected, and the water quality conditions those resources require to thrive. The third step is to determine the flow regimes needed to maintain the desired water quality conditions. Using this approach it is possible to identify flow regimes that are required to sustain ecosystem services in estuarine environments.

Managing both water supplies for people and environmental flows to sustain coastal resources is very difficult. Can we have stable, secure, and sufficient water resources for people and still protect estuarine health? To answer this question, many regions are using adaptive management programs to manage freshwater resources. These programs set goals to protect ecosystem resources, identify indicators, and monitor the indicators over time to ensure that the goals are appropriate and resources are protected. Case studies demonstrate that monitoring and research can determine the ecological and socio-economical impacts of altered freshwater inflows so that stakeholders and managers can make well-informed decisions to manage freshwater inflows to local coasts wisely.

# Acknowledgments

*When you drink the water, remember the spring*—Chinese Proverb

*Then said he unto me, These waters issue out toward the east country, and go down into the desert, and go into the sea: which being brought forth into the sea, the waters shall be healed*—Ezekiel 47:8

This book is based on work that was partially supported by a number of organizations and funding agencies including: the Harte Research Institute for Gulf of Mexico Studies at Texas A&M University-Corpus Christi; the Korea Maritime Institute; the National Oceanic and Atmospheric Administration, CAMEO award NA09NMF4720179; Environmental Protection Agency, grant agreement MX954526; and the Texas Water Development Board, which has provided grant support continuously since 1986 for freshwater inflow studies that led to sampling and analyses that form the basis of many of the approaches described here.

This book is inspired by collaborations with many individuals, but three in particular bear mentioning here. In 1986, Gary Powell (then director of the Bay and Estuary Program, Texas Water Development Board, Austin, TX, USA) asked a seemingly simple question: how much water has to flow into San Antonio Bay to maintain productivity? At the time, it appeared to be a simple question that would be easy to answer within two or three years by measuring processes along salinity gradients and comparing them to changes in flow rates over time. So, there commenced a long period where the focus was on measuring various ecosystem components and discovering that the correlations with inflow were none-existent or so noisy as to be useless. Looking back, it was a rather naïve assumption, and it actually took nearly 20 years to figure out how to reframe the question in a way that it could actually be answered. My friend and colleague, David Maidment (Professor, Civil Engineering, University of Texas at Austin) provided a key insight. We were at a freshwater inflow workshop in 2005 discussing how difficult it was to connect inflow with biological responses when he suggested that we might simply use the water quality paradigm developed by the US Environmental Protection Agency and the Texas Commission on Environmental Quality in analogy to water quantity. This key insight provided a new path of inquiry whereupon the focus changed to characterizing environmental conditions and the response of ecological health to those conditions. Over the decades, key insights into the "Rube Goldberg" nature of water systems and physical processes were provided by George Ward (Senior Research Scientist, Center for Research in Water Resources, University of Texas at Austin). But more importantly, being an engineer, George continually challenged the idea that an ecosystem could be "healthy." With his rapier wit and dry sense of humor, George provided the last piece of the puzzle by forcing us to continually refine, rescope, and rejustify every conclusion we made about the health of the bay or estuary.

# Contents

# Chapter 1
# Introduction

Water development projects are constructed worldwide for many purposes, but most commonly for water supply, flood control, and energy production. Projects always alter hydrology, which means that environmental flow regimes are also altered.

There are three types of environmental flow: instream, inflow, and outflow. Instream flows occur within rivers and streams, and much has been written about the environmental consequences of altered instream flows (Postel and Richter 2003). Inflows occur when rivers flow into the coastal zone, which dilutes marine water and creates brackish estuary habitat. Outflows occur when rivers or estuaries flow out onto continental shelves or nearshore coastal ocean zones.

The purpose of this book is to describe what is currently known about the ecological effects of altered hydrological cycles on coastal and marine areas. It is important to define potential impacts of developments and plan for adaptation or mitigation in the coastal zone.

## 1.1 Problem and Need

Demand for freshwater by human populations is large and continues to grow. Half of the world's major cities are within 50 km of the coast, and coastal populations are 2.6 times more dense than those further inland (Crossland et al. 2005). Because of this, water scarcity and limited or reduced access to water are major challenges facing society and limiting economic development in many countries (MEA 2005). As humans further develop technologies for diverting and capturing freshwater from rivers and streams, greater reductions in freshwater delivery to the coastal zone will occur. The amount of water impounded behind dams globally has quadrupled since 1960 and three to six times as much water is held in reservoirs as in natural rivers (Fig. 1.1; MEA 2005). One of the most severe anthropogenic impacts on coastal areas in the near future will likely be through continued interference with hydrology and water flows to the coast (Pringle et al.

P. A. Montagna et al., *Hydrological Changes and Estuarine Dynamics*,
SpringerBriefs in Environmental Science, DOI: 10.1007/978-1-4614-5833-3_1,
© The Author(s) 2013

**Fig. 1.1** Intercepted
continental runoff entrains
3–6 times more water in
reservoirs than exists in
natural rivers (MEA 2005)

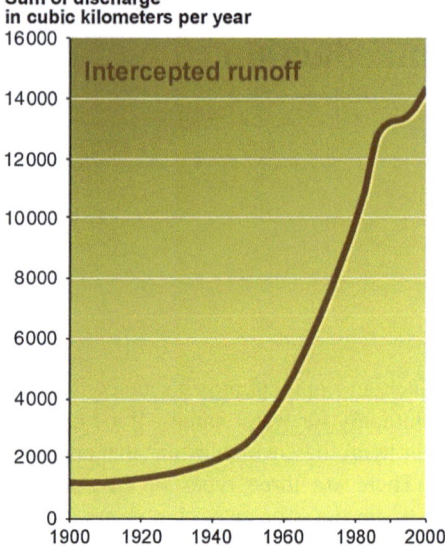

2000). The potential for change in coastal ecosystems due to altered flow regimes
is enormous and a growing concern worldwide.

Two major forces are reshaping freshwater flows to estuaries worldwide: demo-
graphics and engineering. The coastal population is large and continues to grow,
resulting in increasing demand for freshwater. Approximately, 60 % of the people
in the United States live within 60 km of the coast, and 17 of the 20 fastest growing counties are located in coastal areas (Culliton 1998). Freshwater use in the
US has doubled since 1940 and is likely to double again by 2015 (Naiman et al.
1995). Globally, humans use about 54 % of the runoff that is spatially and tempo-
rally available (Postel et al. 1996) and are having a profound effect impact on the
water cycle (Vörosmarty and Sahagian 2000). As the global population continues
to grow, less water will be available for instream flows, inflow to estuaries, or outflows to the coastal zone.

Dams have been constructed throughout human history, but large engineering
marvels are more recent. Large dams were first built in the 1920s through 1930s
to provide hydroelectric power, not water resources. Since then, many large reservoirs have been built to meet an increasing population's needs for water and
energy. Excepting Alaska, the hydrology of nearly every body of freshwater in the
US has been modified by dams, diversions and withdrawals (Naiman et al. 1995),
and similar trends are apparent worldwide (Dynesius and Nilsson 1994; Pringle
et al. 2000). For the first time in human history, these large watershed-scale structures have severely limited inflow to many of the world's estuaries, and consequently altered functioning of these ecosystems (Montagna et al. 2002).

An estuary is defined as a semi-enclosed body of water where salt water
from the ocean mixes with freshwater from rivers and land. Nothing is more

fundamental to the functioning of an estuary than the amount of freshwater delivery to the mixing zone (Dahms 1990; Montagna et al. 2002a). Freshwater inflow regimes vary, but inflows are usually delivered in pulses that arrive in stochastic and complex long-term cycles. The pulses of inflow regimes have four characteristics: frequency, timing, duration, and volume. Altered freshwater inflow has driven changes in coastal ecosystem hydrology, downstream transport of nutrients and sediments, salinity regimes, and has resulted in losses of habitat, biodiversity, and productivity (Montagna and Kalke 1992; Longley 1994; Atrill et al. 1996; Mannino and Montagna 1997; Montagna et al. 2002b; Tolley et al. 2006). Maintaining the hydrological regime and natural variability of an estuary is necessary to maintain its ecological characteristics, including biodiversity.

Because freshwater inflow to estuaries is a major influence on coastal ecosystems, it is important to understand the effects caused by altered freshwater inflow and to create effective management strategies for water resource development and coastal resource management. International attention has become focused on the importance of preserving freshwater flows and the need to develop and employ standards on limitations to the reduction or alteration of flows (Istanbul Water Guide 2009). The European Union (EU) has undertaken several initiatives in recent years, the most important being the European Water Framework Directive (2000/60/EC), which aims to achieve "good ecological status" for all inland and coastal waters by 2015 through the establishment of environmental objectives and ecological targets for surface waters (WFD 2000). The South African National Water Act of 1998 requires that, for any given water resource, sufficient water be set aside to provide for basic human needs and the protection and maintenance of aquatic ecosystems (Republic of South Africa 1998; Thompson 2006). The National Water Policy of India (2002) directs that minimum flow should be ensured in perennial streams for maintaining ecological and social considerations. Within the US, states with large coastal populations (e.g., California, Florida, and Texas) were among the first to face the issue of environmental flows by passing legislation to protect coastal species and resources (Montagna et al. 2002a). This international attention indicates that water shortages, and the consequent reductions of environmental flows, are emerging global issues.

Climate change threatens to change precipitation and temperature patterns in vast regions of the globe. Even with no change in precipitation, increased temperature will increase evapotranspiration, thus creating water deficits in many regions. Although dewatering of estuaries at the current time is driven largely by coastal development and human demand for freshwater, current water management practices may not be adequate to cope with the impacts of climate change. Despite the uncertainty associated with global climate models, the tendency toward more widespread drought increases concomitantly for many arid and semi-arid regions of the globe, including the African Sahel and southern Africa, Central America, the Mediterranean basin, western USA, southern Asia, eastern Australia, and northeastern Brazil (Bates et al. 2008). One immediate threat of reduced precipitation is food security, which depends on irrigation. However, the greater water deficits will lead to greater dewatering of the coastal zone. If river discharge decreases, salinity of coastal ecosystems will increase and the amount of sediment and nutrient delivery will decrease, thereby altering the zonation of plant and animal species as well as the availability of freshwater for human use (Bates et al. 2008; Pollack et al. 2009).

Given the unprecedented change in the water cycle caused by human and climate systems, there are clear needs to manage water resources in the coastal zone using an ecosystem-based approach to protect human health and well-being by sustaining coastal resources. Considerable scientific information is needed to manage coastal ecosystems, such as: What affect will altered freshwater inflow have on coastal resources? What are the relative magnitudes of effects driven by human activities versus climate change? The focus of management initiatives must shift to land planning efforts that conserve water, prevent polluted runoff and groundwater contamination, restore the physical integrity in aquatic ecosystems by increasing natural flow regimes, and promote and protect ecosystem services that could potentially be produced (Ruhl et al. 2003). Despite the growing consensus that the key to maintaining healthy aquatic ecosystems and the services that they provide is to preserve or restore some semblance of a natural flow regime to protect the native flora and fauna, we have continued to implement a piecemeal policy approach making such efforts exceedingly difficult (Katz 2006). The issues of what to do about environmental flows will increase, in importance, worldwide as developing nations further develop water resources for cities, irrigation, and industry. Creating answers to the above questions will provide policy makers and resource managers with science-based ecosystem information, and an array of options to manage environmental flows and water quantities.

# Chapter 2
# Conceptual Model of Estuary Ecosystems

## 2.1 Estuaries

An estuary is a semienclosed coastal body of water, which has a free connection with the open sea and within which, sea water is measurably diluted with fresh water from land drainage (Pritchard 1967). Most estuaries have a series of landscape subcomponents: a river (or fresh water) source, a tidal-estuarine segment, marshes (or mangroves depending on latitude), bays, and a pass (or inlet) to the sea. However, all estuaries are quite different; the landscape of each subcomponent can vary, combinations and connections of these subcomponents can vary, and some subcomponents can be missing. The interaction of three primary natural forces causes estuaries to be unique and different:

- Climate—causing variability in the freshwater runoff and evaporation regimes.
- Continental geology—causing variability in elevation, drainage patterns, landscapes, and seascapes.
- Tidal regime—causing differences in the degree of mixing and elevation of the mixing zone.

Because each of these three physical drivers can vary in a large number of ways, it is easy to imagine how the various combinations of these forces can combine to create a vast array of estuarine typologies. Further variability in estuarine typology is caused by the interactions of these physical drivers.

The physical differences among estuaries are the key to predicting the effects of fresh water alterations. Thus, classifying estuarine typologies is an important first step toward understanding the need for riparian connections to the sea. In spite of the unique signatures of most estuaries, several classification schemes have been presented (Pritchard 1967; Davies 1973; Day et al. 1989).

Based on geomorphology, Pritchard (1952) recognized four estuary typologies: (1) drowned river valleys created by sea level change or sediment starvation in coastal plains, (2) fjords formed by glaciations, (3) bar-built estuaries formed by sediment deposition by winds and tides, and (4) tectonic estuaries caused by faults in the coastal zone. Davies (1973) recognized that there is a continuum of inlet types based on the energy expended on the coast by waves. On one end of the spectrum are lagoons that are enclosed by sandy spits and at the other end of the spectrum are deltas that are muddy and formed by river processes. Day et al. (1989) recognized that all previous definitions still do not encompass all estuarine typologies and suggested that an estuary is any coastal indentation that remains open to the sea at least intermittently and has any amount of freshwater inflow at least seasonally.

Water balance is the second important defining characteristic of estuaries. The freshwater balance is simply the sum of the water sources minus the sum of the water losses. The many sources of fresh water to the coastal zone include: rivers, streams, groundwater, direct precipitation, point-source discharges, and non-point-source runoff. There are fewer mechanisms that cause losses of fresh water, but these primarily include evaporation and freshwater diversions for human use. Pritchard (1952) recognized three classes of estuaries based on natural hydrological processes: (1) positive estuaries where freshwater input from rain, runoff, rivers, and groundwater exceeds evaporation; (2) neutral estuaries where the sources and losses are in balance; and (3) negative or inverse estuaries where evaporation exceeds the combined sources of fresh water. Depending on climate, some systems change seasonally, being positive during rainy seasons and negative during dry seasons. Many estuaries in the world have strong year-to-year variability caused by interannual climatic variability.

## 2.2  Human Interactions

Human activities and water resource development can change the freshwater balance in estuaries dramatically (Fig. 2.1). Freshwater diversions used as water supplies for large humans populations or large agricultural areas are large sinks or losses to systems. However, return flows (e.g., wastewater or industrial water) add a source of fresh water to ecosystems. In many cases, the diversions and return flows can be roughly in balance if they are planned as a unit using integrated water planning. But this is rarely, if ever the case. Because many water systems depend on gravity feeds to save pumping expenses, diversions are often taken upstream and returns (minus losses to leaks and use) are put in downstream. Depending on intervening elevation and geomorphology, return flows can even be put into different watersheds. When the demand for water is large relative to the supply, the water balance can be altered significantly.

Clearly, the estuaries most at risk from human activities are those that already have a negative water balance throughout the year or during certain seasons or times. Those estuaries that are neutral but have large upstream water demands are also at great risk of degradation due to altered flow regimes. The change of fresh water volume will have profound effects on salinity in a shallow estuary (e.g., coastal plain estuaries or lagoons), but a smaller effect on a deeper estuary

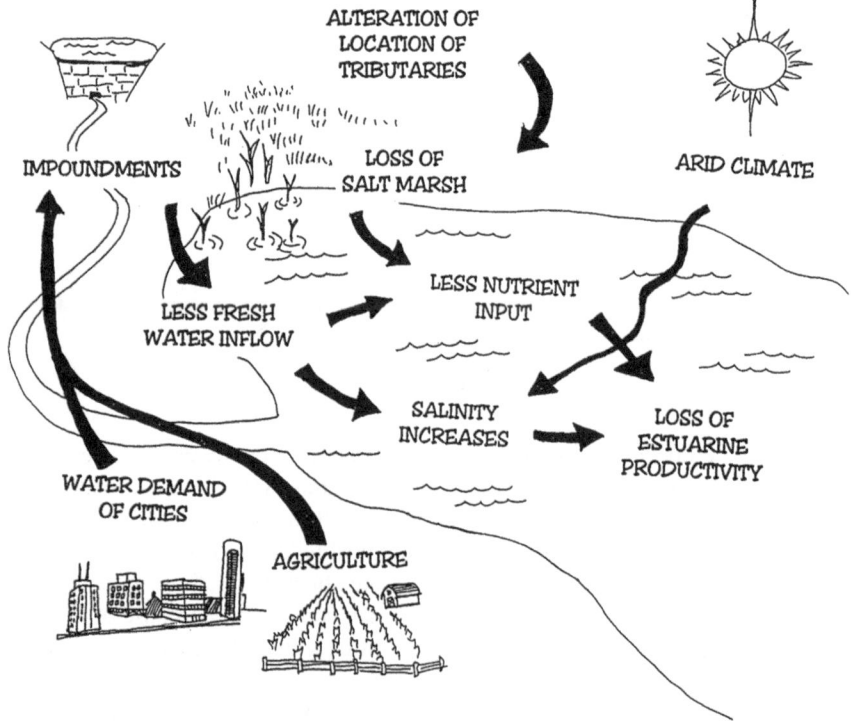

ALTERATION OF
LOCATION OF
TRIBUTARIES

IMPOUNDMENTS

LOSS OF
SALT MARSH

ARID CLIMATE

LESS FRESH
WATER INFLOW

LESS NUTRIENT
INPUT

SALINITY
INCREASES

LOSS OF
ESTUARINE
PRODUCTIVITY

WATER DEMAND
OF CITIES

AGRICULTURE

**Fig. 2.1**  Effects of altered inflow on estuaries (Montagna et al. 1996)

(e.g., fjords or tectonic estuaries). This difference of effect is often caused by shallow estuaries having smaller water volumes than deeper estuaries.

Given that humans can now alter many factors of the water cycle, it is imperative that freshwater resources be managed effectively to protect downstream ecological resources. Beginning in the 1960s, scientists began to investigate how altered freshwater flows to the coast might affect biological resources (Copeland 1966; Hoese 1967). Since then, there have been at least two major compilations of papers on the topic: Cross and Williams (1981) and Montagna et al. (2002a). As a result of these two symposia and other work there have been two important reviews (Alber 2002; Estevez 2002) from which a conceptual model has emerged that helps us to identify inflow effects (Fig. 2.2).

Following a review of the practices in three states (California, Florida, and Texas) where there is a long history of inflow studies, Alber (2002) defined the scientific framework for identifying the effects of inflow on estuarine resources. Historically, all freshwater inflow methodologies started from the perspective of hydrology or resource protection. The earlier approaches were all focused on resources such as protection of fish, charismatic, or iconic species. The problem quickly encountered is that the relationship between biology and hydrology is

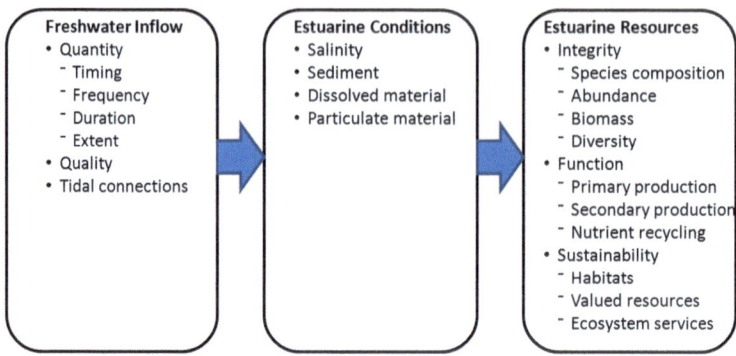

**Fig. 2.2**  Conceptual model of inflow effects (modified from Alber 2002)

complex and embedded in the food web and material flow dynamics of estuaries. For example, one cannot grow fish by simply adding water to a fish tank. These experiences led to a generic framework where inflow hydrology drives estuarine condition and estuarine condition drives biological resources (Fig. 2.2).

Ultimately, biological resources in estuaries are affected by salinity more than flow by itself. Salinity is affected by flow, but there are complexities because of the interactions between tides and geomorphology. Consequently, all salinity-flow relationships are characterized with very high variance or scatter, especially in the low flow end of the spectrum. Because of the links among flow, salinity, and biology, all the resource based approaches are multistep. First, the resource to be protected is identified. Second, the salinity range or requirements of that resource are identified in both space and time. Third, the flow regime needed to support the required distribution of salinity is identified, usually using hydrodynamic and salinity transport models.

The usefulness of the environmental flow framework (Fig. 2.2) is that estuarine resources are categorized into the familiar framework used to describe ecological health (i.e., integrity, function, and sustainability). Two new terms are added: valued resources and ecosystem services. The resources are typically called "valued ecosystem components" or VECs. These are resources that are identified by stakeholders as having esthetic, ecological, economic, or social value. These resources include bioengineers (or foundation species) that create habitat, fisheries species, and birds. These species are typically charismatic, characteristic, or iconic to an area. Ecosystem services are the benefits provided by the environment to human health and well-being (Costanza et al. 1997). It is clearly in the socioeconomic interest to sustain ecosystem services, especially those provided by VEC habitats such as oyster reefs, marshes, and seagrass beds.

Another important feature of the environmental flow conceptual model (Fig. 2.2) is that it is analogous to the well-accepted environmental risk assessment (or risk management) paradigm (Fig. 2.3) that has evolved in water quality management since the 1970s. The risk assessment paradigm is also known as the pressure-state-response (PSR) model. In the water quality PSR model, the pressure is applied from a toxic substance, the state represents the presence or

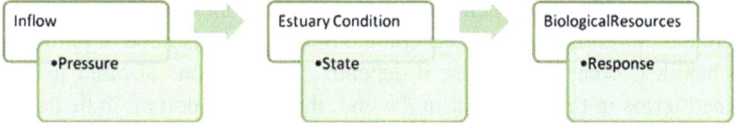

**Fig. 2.3** Relationship between conceptual model of inflow effects and the water quality risk assessment paradigm

concentration of a substance, and response is the biological response to that state. Management actions are another way in which "Response" can be thought of. The analogy here is that flow is the pressure, estuarine condition is the state, and change in estuarine resources is the biological response (Fig. 2.3). This is a very powerful way to think about the effects of inflow on estuarine resources, because it helps us to define the ecological health of estuaries. Assessing risk by defining health is often the first step in managing environmental resources.

---

**Defining Ecological Health**

- Ecological health is assessed by determining if <u>indicators</u> of ecological <u>conditions</u> are in an acceptable range.
- Indicators are measures (or metrics) of ecological health for which sufficient information exists to establish an acceptable range of responses across broad spatial and temporal scales.
- Ecological condition is the status of ecological <u>function</u>, <u>integrity</u>, and <u>sustainability</u>.
- Ecological function is judged acceptable when the ecosystem provides important ecological processes.
- Ecological integrity is acceptable when the ecosystem has a balanced, resilient community of organisms with biological diversity, species composition, structural redundancy, and functional processes comparable to that of natural habitats in the same region.
- Ecological sustainability is acceptable when an ecosystem maintains a desired state of ecological integrity over time.

---

Defining ecological health is a vexing issue. Consider the analogy with human health. Scientists have proven that the normal human body temperature range is 36.4–37.2 °C. If a person's body temperature is above this range then they have a fever, and are likely sick. This example illustrates several important principles about human health as it relates to defining ecological health and how the definition has evolved for water quality assessment (Montagna et al. 2009). It is easy to integrate the conceptual model of inflow effects and the risk assessment paradigm to provide a general reference frame to define "ecological health" (see Box). Indicators of health have to be identified and the indicators must be within an acceptable range. Two difficulties are that there are no simple indicators of ecological health; and when an indicator can be measured, there are seldom sufficient data to know what the acceptable ranges are. Also, the definition of ecological health is complex because it depends on definitions of other terms (those underlined terms in the box). But in the end, the most important indicator is likely ecological sustainability. Sustainability is the ultimate definition of ecological health because an environment that is sustainable is healthy in the strict sense.

## 2.3 Hydrology and the Water Cycle

Water is the most widely used natural resource on the Earth. However, less than 1 % (0.7 %) of the water on the Earth is fresh and of sufficient quality to be classified as drinkable. Only two-one thousandths of 1 % (0.00002) is readily available in streams and lakes for humans to use to drink, bathe, or irrigate crops. The same amount of water is available today as 2,000 years ago, yet the world's population was just 3 % of what it is today, thus water availability is an extreme limit to growth and prosperity (Lane et al. 2003).

The Earth is often referred to as the blue planet because water covers about two-thirds of its surface. Because water is so plentiful on the Earth, the water cycle influences most climatic and surface geologic processes. Two dominant processes drive the water cycle: evapotranspiration and precipitation (Fig. 2.4). Water resource planners, however, appear to be concerned mainly with precipitation because they can manipulate runoff. Rain over large areas interacts with land elevation to form drainage patterns and familiar landscapes, e.g., tributaries, streams, rivers, and wetlands. These drainage systems are watersheds. If the

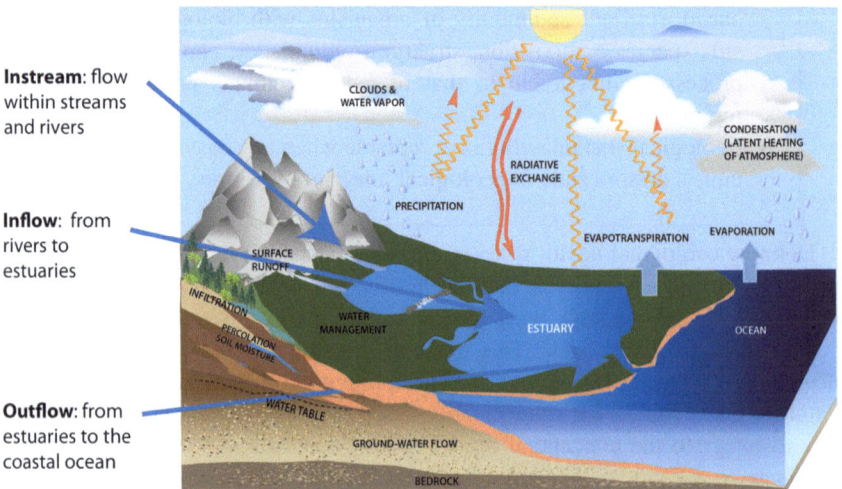

**Fig. 2.4** Locations of environmental flows in the water cycle

watershed is adjacent to the coastal zone, then the ultimate drainage basin is the estuary, where fresh water mixes with sea water. Freshwater runoff can, and has been, manipulated to develop hydroelectric power or create reservoirs for water storage. Altering watercourses alters riparian, wetland, and estuarine habitats (Fig. 2.1). In coastal regions, the effect of freshwater inflow alteration depends on the type of estuary, the biological components present, and the climatic setting.

It is important to be able to build water budgets in order to manage flow to estuaries. In most places in the World, there are already abundant data on water flow and availability because of the importance of water to agriculture, cities, and industry. However, rainfall and river flow alone are not always sufficient to calculate total inflow to estuaries. Fresh water can also enter estuaries via runoff from land and through seepage of ground water. The degree to which these are important will likely be unique to each system. Initially, runoff and ground water seepage can be assumed to be small or insignificant, but eventually estimates of these inputs will be needed. Runoff is usually calculated based on models of land elevation, drainage patterns, and rainfall. Unlike surface water, ground water is difficult to observe and measure. Runoff estimates are found commonly, but groundwater inputs are rarely used in water cycle planning budgets.

Evapotranspiration is the water loss from direct evaporation of the water surface and water lost through plants. Water loss from a system due to evapotranspiration must also be known, especially in hot, dry areas where the volumes can be large. Evaporation can be measured directly by placing water in a pan and measuring the volume lost daily. Total water loss can be calculated as the product of the evaporation rate and the surface area of the water body.

Human activities and water use and reuse must also be accounted for. Water authorities usually record the amount of water withdrawn from ecosystems for human use, thus it should be relatively easy to obtain this information over a long period of time. Water is also returned to the environment after use. This is usually in the form of wastewater, but sometimes it is agricultural runoff, or industrial cooling water. These quantities should also be known and accounted for in determining the total inflows to estuaries.

Once all the basic parameters are known, the water balance can be calculated. The water balance for an estuary is simply the sum of the inputs, minus the sum of the outputs:

$$Water\ balance = (precipitation + river\ flows + runoff + return\ flows)$$
$$- (evaporation + diversions)$$

In this case, precipitation is directly falling on the surface of the estuary. This value will be small and insignificant in drowned river valleys, but large and significant in large coastal bays.

The natural hydrological parameters (i.e., rain, inflow, runoff, and evaporation) are driven by climate, which varies considerably within years (i.e., seasons) and among years. It is clear from the recent debate on climate change that different climatic cycles range from decades to millennia. The natural variability in river flow and levels is essential information to know for water resource management.

In the end, environmental flows are likely most beneficial if they mimic the natural flow regime. The regime is composed of several characteristics including the variability of flow rates and levels. Storms create floods, which in turn create events that drive change in ecosystems. The size of these events (or disturbances) has several important characteristics: frequency, duration, extent, and timing. Each of these characteristics will have a statistical distribution with a mean and variance. Frequency is how often flood events occur. Duration is how long the event lasts. Extent is the magnitude or size of the event. Timing is the seasonality of events. For example, during a recent demonstration project to divert flow back into a marsh restoration area, the frequency, timing, and duration of floods was restored; but the extent was not because large volumes of water were trapped behind a dam (Ward et al. 2002). This project demonstrates that all characteristics of freshwater inflow to estuaries have to be characterized so that the total regime can be understood.

## 2.4  Tides and Residence Times

In all estuaries, the dilution of seawater by the volume of fresh water inflow is affected by the amount of salt water in the system. The volume of salt water is controlled by two main factors: tides and the volume of the receiving body.

Tides are the rise and fall of the sea around the edge of the land. Tides are driven by the gravity of the Moon and Sun, which causes a bulge in the water on the Earth's surface. Far out at sea, tidal changes go unnoticed, but are very important to the plants and animals that live on the edge of sea, in the "intertidal zone." Tides occur in primarily three different patterns. Diurnal tides are where there is one high and one low tide each day. Semidiurnal tides are where the rise and fall is repeated twice each day. Semidiurnal mixed tides occur where there are unequal tidal heights each day. Tidal levels are different in different parts of the world, but in general the tidal ranges can be small (<1 m) or great (>3 m). Thus, tidal range can be categorized as microtidal, mesotidal, or macrotidal.

The combination of river flow and tides mean there are different types of circulation patterns in estuaries. Salt wedge estuaries are where river flow dominates the mixing and typically freshwater overflows salty water giving rise to strong stratification at a specific point. Well-mixed estuaries are where the wind dominates mixing and there is a strong salinity gradient from the river to the sea, but little to no stratification of the water column. Partially mixed estuaries are where tides dominate the mixing patterns so that there is stratification of salt water on the bottom and fresher water on the top, but this gradient can be variable along the axis of the estuary. Fjord-type estuaries are where sea water is trapped in deep parts of the bay, typically behind a sill, and there is little to no exchange with the surface water.

Because of the mixing of salt and freshwater, water budgets and salt budgets are very important to understanding flow dynamics. In a water budget, the total volume transported out in a unit of time is the sum of the volume transported in by tides plus the volume of the freshwater inflow transported into the estuary from the river. Thus, the residence time (or flushing time) is the volume of the estuary ($V_{estuary}$) divided by the rate of flow of water leaving the estuary ($T_{out}$), i.e., Flushing Time = $V_{estuary}/T_{out}$. The flushing time is very important because it controls the carrying capacity of wastes, the flushing time for fresh water, the flushing by tidal action, the effects of mixing, and it can be affected by coastal upwelling and downwelling. Thus, flushing rate is the master variable that controls nearly all estuarine processes.

## 2.5 Estuarine Condition and Water Column Effects

Watershed development such as the construction of dams and withdrawal of water for human use has changed flow regimes, transport of sediments and nutrients, modified habitat, and disrupted migration routes of aquatic species (MEA 2005). These modifications to the hydrologic cycle affect the quantity, quality, and timing of freshwater inflows, and the health of estuaries. Understanding the cascading link between inflow, condition, and response (Fig. 2.1) is the key to understanding how change driven by human and climate systems can drive resistance and resilience of biological communities.

### 2.5.1 Salinity

The salinity at any point within an estuary reflects the degree to which seawater has been diluted by freshwater inflows. Estuaries are transitional zones between freshwater and marine environments, and as such, display gradients of salinity (0 in freshwater to 35 ppt in seawater) and nutrients (high in freshwater, low in seawater; Montagna et al. 2010). When less dense freshwater flows into more dense saltwater, the freshwater has a tendency to remain primarily on the surface layer (Kjerfve 1979). However, winds and tides tend to mix the water column, creating longitudinal and vertical salinity gradients within estuaries (Day et al. 1989). Estuaries can be classified based on their water balance: (a) positive estuaries have freshwater inputs that exceed evaporation, (b) neutral estuaries have a balance between freshwater input and evaporation, and (c) negative estuaries have evaporation that exceeds freshwater input (Pritchard 1952). Depending on the hydrologic cycle, a system may change seasonally from being a positive to a negative estuary, or vice versa.

Water development projects can reduce the delivery of freshwater to estuaries and also affect the timing of inflow pulses, which can affect organisms adapted to the original salinity conditions. Although estuarine organisms generally have a wide salinity tolerance (euryhaline), most are located only within a portion of

**Table 2.1** Selected references for salinity effects on estuarine macrobenthic and epibenthic organisms

| Authors | Organism(s) studied | Study location | Salinity tolerance results |
|---|---|---|---|
| Chadwick and Feminella (2001) | Burrowing mayfly *Hexagenia limbata* | USA (Alabama) | Laboratory bioassays showed that *H. limbata* nymphs could survive elevated salinities (LC50 of 6.3 ppt at 18 °C, 2.4 ppt at 28 °C). Similar growth rates at 0, 2, 4, and 8 ppt |
| Saoud and Davis (2003) | Juvenile brown shrimp *Farfantepenaeus aztecus* | USA (Alabama) | Growth significantly higher at salinities of 8 and 12 ppt than at salinities of 2 and 4 ppt |
| Tolley et al. (2006) | Oyster reef communities of decapod crustaceans and fish | USA (Florida) | Upper stations (~20 ppt) and stations near high-flow tributaries (6–12 m$^3$ s$^{-1}$) were typified by decapod *Eurypanopeus depressus* and gobiid fishes. Downstream stations (~30 ppt) and stations near low-flow tributaries (0.2–2 m$^3$ s$^{-1}$) were typified by decapods *E* |
| Montagna et al. (2008a) | Southwest Florida mollusk communities | USA (Florida) | *Corbicula fluminea*, *Rangia cuneata*, and *Neritina usnea* only species to occur <1 psu. *Rangia cuneata* good indicator of mesohaline salinity zones with tolerance to 20 psu. Gastropod *N. usnea* common in fresh to brackish salinities. *Polymesoda caroliniana* present between 1–20 psu (oligo- to mesohaline zones). *Tagelusplebius, Crassostrea virginica, Mulinia lateralis, Littoraria irrorata,* & *Ischadium recurvum* good indicators for polyhaline salinity zones. |
| Montague and Ley (1993) | Submersed vegetation and benthic animals | USA (Florida) | Mean salinity ranged from ~11 to 31 ppt. Standard deviation of salinity was best environmental correlate of mean plant biomass and benthic animal diversity. Less biota at stations with greater fluctuations in salinity. For every 3 ppt increase in standard |
| Rozas et al. (2005) | Estuarine macrobenthic community | USA (Louisiana) | Increased density and biomass with increases in freshwater inflow and reduced salinities. Salinity ranged from 1 to 13 psu |
| Finney (1979) | Harpacticoid copepods *Tigriopus japonicus, Tachidius brevicornis, Tisbe sp.* | USA (Maryland) | All species tested for response to salinities from 0 to 210 ppt. *Tigriopus* became dormant at 90 ppt died at 150 ppt. *Tachidius* became dormant at 60 ppt, died at 150 ppt. *Tisbe* died shortly after exposure to 45 ppt |

(continued)

**Table 2.1** (continued)

| Authors | Organism(s) studied | Study location | Salinity tolerance results |
|---|---|---|---|
| Kalke and Montagna (1991) | Estuarine macrobenthic community | USA (Texas) | Chironomid larvae and polychaete *Hobsonia florida*: increased densities after freshwater inflow event (1–5 ppt). Mollusks *Mulinia lateralis and Macoma mitchelli*: increased densities and abundance during low flow event (~20 ppt). *Streblospio benedicti and Medioma* |
| Keiser and Aldrich (1973) | Postlarval brown shrimp *Penaeus aztecus* | USA (Texas) | Shrimp selected for salinities between 5 and 20 ppt |
| Montagna et al. (2002b) | Estuarine macrobenthic community | USA (Texas) | Macrofauna increased abundances, biomass and diversity with increased inflow; decreased during hypersaline conditions. Macrofaunal biomass and diversity had nonlinear bell-shaped relationship with salinity: maximum biomass at ~19 ppt |
| Zein-Eldin (1963) | Postlarval brown shrimp *Penaeus aztecus* | USA (Texas) | In laboratory experiments with temperatures 24.5–26.0 °C, postlarvae grew equally well in salinities of 2–40 ppt |
| Zein-Eldin and Aldrich (1965) | Postlarval brown shrimp *Penaeus aztecus* | USA (Texas) | In laboratory experiments with temperatures <15 °C, postlarval survival decreased in salinities <5 ppt |
| Allan et al. (2006) | Caridean shrimp *Palaemon peringueyi* | South Africa | At constant salinity of 35 ppt, respiration rate increased with increased temperature. At constant temperature of 15 °C, respiration rate increased with increased salinity |
| Ferraris et al. (1994) | Snapping shrimp *Alpheus viridari*, Polychaete *Terebellides parva*, sipunculan *Golfingia cylindrata* | Belize | Organisms subjected to acute, repeated exposure to 25, 35, or 45 ppt. *Alpheus viridari* hyperosmotic conformer at decreased salinity, but osmoconformer at increased salinity. *Golfingia cylindrata* always osmoconformer. *Terebellides parva* always osmoconformer; decreased survival |
| Lercari et al. (2002) | Sandy beach macrobenthic community | Uruguay | Abundance, biomass, species richness, diversity, and evenness significantly increased from salinity of ~6 ppt to salinity of ~25 ppt |
| Chollett and Bone (2007) | Estuarine macrobenthic community | Venezuela | Immediately after heavy rainfall (~25 psu), spionid polychaetes showed large increases in density and richness versus normal values (~41 psu) |

(continued)

**Table 2.1** (continued)

| Authors | Organism(s) studied | Study location | Salinity tolerance results |
|---|---|---|---|
| Dahms (1990) | Harpacticoid copepod *Paramphiascella fulvofasciata* | Germany (Helgoland) | After 2 h, no mortality in salinities of 25–55 ppt. Almost all displayed dormant behavior <20 and >55 ppt |
| McLeod and Wing (2008) | Bivalves *Austrovenus stutchburyi* and *Paphies australis* | New Zealand | Sustained exposure (>30 d) to salinity <10 ppt significantly decreased survivorship |
| Rutger and Wing (2006) | Esturaine macroinfaunal community | New Zealandz | Infaunal community in low salinity regions (2–4 ppt) showed low species richness and abundance of bivalves, decapods, and Orbiniid polychaetes, but high abundance of amphipods and Nereid polychaetes compared to higher salinity regions (12–32 ppt) |
| Drake et al. (2002) | Estuarine macrobenthic community | Spain | Species richness, abundance, and biomass decreased in the upstream direction, positively correlated with salinity. Highly significant spatial variation in macrofaunal communities along the salinity gradient. Salinity range: 0–40 ppt |
| Normant and Lamprecht (2006) | Benthic amphipod *Gammarus oceanicus* | Baltic Sea | Low salinity basin (5–7 psu). Physiological performance examined from 5 to 30 psu. Feeding and metabolic rates decreased with increasing salinity; nutritive absorption increased. Feces production and ammonia excretion rates decreased strongly from lowest to highest salinity. Greatest scope for growth at 7 psu. |

their salinity range. Thus, salinity gradients play a major role in determining the distribution of estuarine organisms (Table 2.1). Secondary production by estuarine benthic macrofauna in particular is known to increase with increases in freshwater inflow (Montagna and Kalke 1992). Salinity gradients also can act as barriers to predators and disease. Two important oyster predators in Gulf of Mexico estuaries, the southern oyster drill *Thais haemastoma* and the stone crab *Menippe mercenaria* are intolerant of sustained salinities below 15 psu (Menzel et al. 1958; MacKenzie 1977). Freshwater inflow, depending on the volume, can dilute or even eliminate infective *Perkinsus marinus* oyster disease particles in low salinity areas (Mackin 1956; La Peyre et al. 2009). The timing of freshwater inflows is also important to estuarine organism abundance and distribution because the organisms have evolved over long periods to particular regimes of freshwater inflow and associated hydrologic conditions (Montagna et al. 2002).

## 2.5.2  Sediments

In addition to changing salinity levels, freshwater inflow provides nutrients, sediments, and organic materials that are important for overall productivity of the estuary. Thus, any upstream changes in inflow will affect the amount and timing of their delivery to the estuary as well (Alber 2002). High estuarine turbidity is generally observed during high-flow periods due to elevated sediment inputs. Sediments are delivered to estuaries from rivers and streams by freshwater inflow, which helps to build and stabilize wetlands, tidal flats, and shoals (Olsen et al. 2007). Particulate matter carried by rivers also provides the primary energy source for organisms living in the estuarine environment (Day et al. 1989).

Freshwater diversion from estuaries is decreasing the delivery of water and sediment to the coastal zone. Within the continental US, approximately 90 % of the sediment being eroded from land is stored somewhere between the river and the sea (Meade et al. 1990). Changes in sediment discharge over the past 200 years are primarily due to anthropogenic factors including (a) deforestation and agriculture, (b) changes in land management strategy, and (c) construction of dams, diversions and levees (McKee and Baskaran 1999). Worldwide, reservoirs and water diversions have resulted in a net reduction of sediment delivery to estuaries by roughly 10 %, and prevent about 30 % of sediments from reaching the oceans (Syvitski et al. 2005; Vörösmarty et al. 2003).

## 2.5.3  Nutrients

The nutrient content of freshwater flows entering estuarine waters is important because it is closely linked to primary production (Valiela 1995). In estuarine systems, nitrogen is the principal limiting element, followed by phosphorus.

The addition of nutrients to estuaries is a natural process that has been greatly enhanced by human activities. In recent decades, population growth, agricultural practices, wastewater treatment plants, urban runoff, and the burning of fossil fuels have greatly increased nutrient inputs over the levels that occur naturally (Bricker et al. 1999). The concentrations of nutrients in estuaries are dynamic in space and time as a function of inputs and outputs from river flows and oceanic exchange as well as biological uptake and regeneration (Day et al. 1989). Salinity is generally an inverse indicator of the availability of land-derived nutrients, with low salinities (high freshwater inflow) linked to high nutrient concentrations (Pollack et al. 2009; Montagna et al. 2010).

Freshwater inflow can enrich estuarine nutrients and increase primary and secondary production (Livingston et al. 1997; Brock 2001). Conversely, decreased inflow has been linked to decreased rates of both primary and secondary production (Drinkwater and Frank 1994). Excess loading of nutrients to coastal waters can cause dense, long-lived algal blooms that block sunlight to submerged aquatic vegetation. The decay of these blooms consumes oxygen that was once available to fish and shellfish, which can result in anoxic or hypoxic conditions (Rabalais and Nixon 2002). Excess nutrients can thus cause degraded water quality and affect the use of estuarine resources such as fishing, swimming, and boating (Bricker et al. 1999).

## 2.5.4  Biological Indicators

Change in freshwater inflow to an estuary not only changes the salinity of an estuary, but also nutrient concentrations. Increases in freshwater inflow usually lead to an increase in bioavailable nutrients, which in turn stimulate primary production. This primary production is often in the form of phytoplankton growth. The phytoplankton growth in turn stimulates secondary production by organisms such as zooplankton and benthic suspension feeders. Following the increase in secondary production, there is often an increase in tertiary production by organisms such as shrimps and fishes. This process is an oversimplification of the biological response to an increase in freshwater inflows. Every estuary and coastal zone is different and complex food webs exist rather than simple food chains. Because it is impossible to determine the exact changes in every population as they respond to changes in freshwater inflow, we instead approximate biological effects using biological indicators.

Biological indicators are individual species or communities of species that integrate changes in the environment so that when monitored, can indicate changes or stability in a particular environment. We expect indicator organisms to do for us today what canaries did for coal miners in the eighteenth and nineteenth centuries. Indicator organisms should have at least five characteristics that make them useful in applied research (Soule 1988). (1) They should direct our attention to qualities of the environment. (2) They should give us a sign that some characteristic is present. (3) They should express a generalization about the environment. (4) They should suggest a cause, outcome or remedy. (5) Finally, they should show a need for action.

Benthic invertebrate communities have been widely used as indicators of ecological health in environmental assessment, pollution detection, and ecological monitoring studies. The benthic community is unique among coastal and marine organisms for several reasons. First, they are predominantly permanent residents of estuaries, unlike much of the more visible nekton that are made up of large populations of migratory organisms. Second, they are relatively long-lived compared to plankton. Third, the benthos are relatively immobile and fixed in space, unlike nekton and plankton that move freely or with currents. In addition, everything dies and ends up in the detrital food chain, which is utilized by the benthos. Because of gravity, there is a record of all environmental change in the sediments, and benthos are commonly referred to as the "memory" of the ecosystem because this record of past events is layered in the sediments. This combination of characteristics means that the benthic community integrates change in ecosystems over long time scales. Benthos are therefore the best sentinel group responding to changes in external conditions without the complication of movement to different regions of the coastal zone. Because benthic organisms are relatively immobile, they are usually the first organisms affected by environmental stress. Many ecological monitoring programs use benthic abundance, biomass, and diversity as ecological indicators of the state, productivity, or health with respect to changes in the environment.

Diverse and abundant populations of benthic invertebrates provide a necessary food source for many aquatic and terrestrial species. Because of the importance of benthic organisms in the estuarine food chain, fluctuations in their abundance can influence recruitment patterns in coastal fisheries and avian migratory behavior. Therefore, it is important to continuously monitor the abundance and diversity of benthic infauna within an estuarine system.

There are good ecological conceptual models that provide a scientific basis for interpreting the data generated in benthic monitoring and detection studies. These approaches utilize many single species, community studies, and statistical models. One of the most important concepts is the succession model proposed by Rhoads et al. (1978). They applied scientific theories of ecological succession and its relation to productivity to suggest ways that dredge-spoil could be managed to enhance productivity. The same year, Pearson and Rosenberg (1978) published a review showing how benthic community succession changed in relation to organic enrichment. The central tenant of this theory is that distance from a pollution source is analogous to time since a natural disturbance. Thus, the sequence of colonization and succession events that occur after a disturbance are similar to the changes in communities observed with distance from a pollution source. There is typically a gradient from smaller, less diverse, pioneering species limited to surface sediments to larger, more diverse, climax assemblages of deeper dwelling organisms. The gradient changes over both distance from a pollution source or is represented by community development over time after a disturbance. Thus, we have a scientific justification for benthic community structure and biodiversity studies as an assessment tool. Since these two classic studies, numerous other studies have demonstrated the value of benthic communities as an excellent indicator of environmental health.

Ecological health can be defined by benthic metrics employing the following series of linked definitions: The condition of integrity is assessed when community structure and diversity are stable over long periods of time. The condition of function is assessed when biomass (the best indicator of productivity) is stable over long time periods. These three metrics: abundance, biomass, and species diversity are easily obtainable in routine sampling programs.

An important objective of many resource agencies is to quantify the relationship between bioindicators of marine resource populations and freshwater inflows to bays and estuaries. However, there is year-to-year variability in population densities and successional events in estuarine communities. This year-to-year variability is apparently driven by long-term, and global-scale climatic events. For example, El Niño affects rates of precipitation and concomitantly rates of freshwater inflow along the Texas coast, which in turn influences salinity patterns in Texas bays (Tolan 2007). Therefore, the best approach is to document long-term changes in populations and communities that are influenced by freshwater inflow. The best indicator of productivity is the change in biomass of the community over time (Banse and Moser 1980). Based on initial sampling of 1–4 years of benthic data in Texas bays, it was originally concluded that inflow does increase benthic productivity (Kalke and Montagna 1991; Montagna and Kalke 1992, 1995). However, further analysis of the data set over a 5-year period demonstrated that the largest effect may not be on productivity, but may be on community structure (Montagna and Li 2010). This implies that reduced inflows may not only reduce productivity (a measure of ecosystem function), but may also change the composition of species in an estuary (a measure of ecosystem structure).

Texas Coastal Bend estuaries were studied over a 20-year period by Montagna (2008) to determine the long-term response of benthic organisms to freshwater inflow. Results show that the biological effects on benthic communities appear to be driven by the El Niño cycle. Flood conditions introduce nutrient rich waters into the estuary that result in lower salinity. During El Niño periods, the lowest salinities and highest nutrient values were recorded. During these periods, the spatial extent of the freshwater fauna is increased, and the estuarine fauna replaced the marine fauna in the lower end of the estuary. The high level of nutrients stimulated a burst of benthic productivity (of predominantly freshwater and estuarine organisms), which lasted about 6 months. This was followed by a transition to a drought period with low inflow resulting in higher salinities, lower nutrients, dominance by marine fauna, decreased productivity, and decreased abundances.

Florida Bay was examined to determine the relationship between commercially important pink shrimp (*Farfantepenaeus duorarum*) recruitment and freshwater inflow characteristics (Browder et al. 2002). Experiments were conducted to determine the rates of juvenile shrimp growth and survival at varying temperatures and salinities and the results were used to refine an existing model of potential pink shrimp recruitment. Results showed high survival over a wide salinity range except at extreme temperatures. In particular, shrimp were least tolerant of high salinity at low temperatures and low salinity at high temperatures. Maintenance of freshwater inflow was important to provide favorable salinities over the greatest amount of suitable and accessible habitat. Timing of flows in relation to arrival of postlarvae from offshore spawning grounds was also found to be important (Browder et al. 2002).

# Chapter 3
# Case Studies

There have been at least two major compilations of research on the topic of
freshwater inflow. There was a symposium convened in 1980 in San Antonio,
Texas to identify the issues regarding "Freshwater Inflow to Estuaries" (Cross and
Williams 1981). The goal of the symposium was to identify potential solutions
and recommendations to deal with the issues of altered inflow regimes. A second
symposium was convened in 2001 in St. Pete Beach, Florida entitled "Freshwater
Inflow: Science, Policy and Management" (Montagna et al. 2002a). The second
symposium is notable because in the intervening 21 years, many agencies began
to implement freshwater inflow rules and regulations, performed research on the
effects of the rules, and even attempted to restore estuaries where inflow was
reduced. One important aspect of nearly all the minimum freshwater inflow rules
is that they are amenable to adaptive management. Results from ongoing monitoring and assessment programs are used to modify and optimize the operating
decisions. This is very important. Following is a brief summary of case studies in
Texas, Florida, California, Australia, South Africa, and Europe.

## 3.1 Texas, USA

### 3.1.1 Physical Background

There are seven major estuarine systems along 600 km of coastline (Fig. 3.1,
Longley 1994). All seven Texas estuaries have similar geomorphic structure and
physiography. Barrier islands are parallel to the mainland along the coast. Between
the islands and the mainland, there are lagoons. The lagoons are interrupted with
drowned river valleys that form the bay and estuarine systems. There are Gulf inlets
through the barrier islands, which connect the sea with the lagoon behind the island.
The lagoon opens to a large primary bay. There is a constriction between the primary bay and the smaller secondary bay. Most bays are fed by just one or two rivers

P. A. Montagna et al., *Hydrological Changes and Estuarine Dynamics*,
SpringerBriefs in Environmental Science, DOI: 10.1007/978-1-4614-5833-3_3,
© The Author(s) 2013

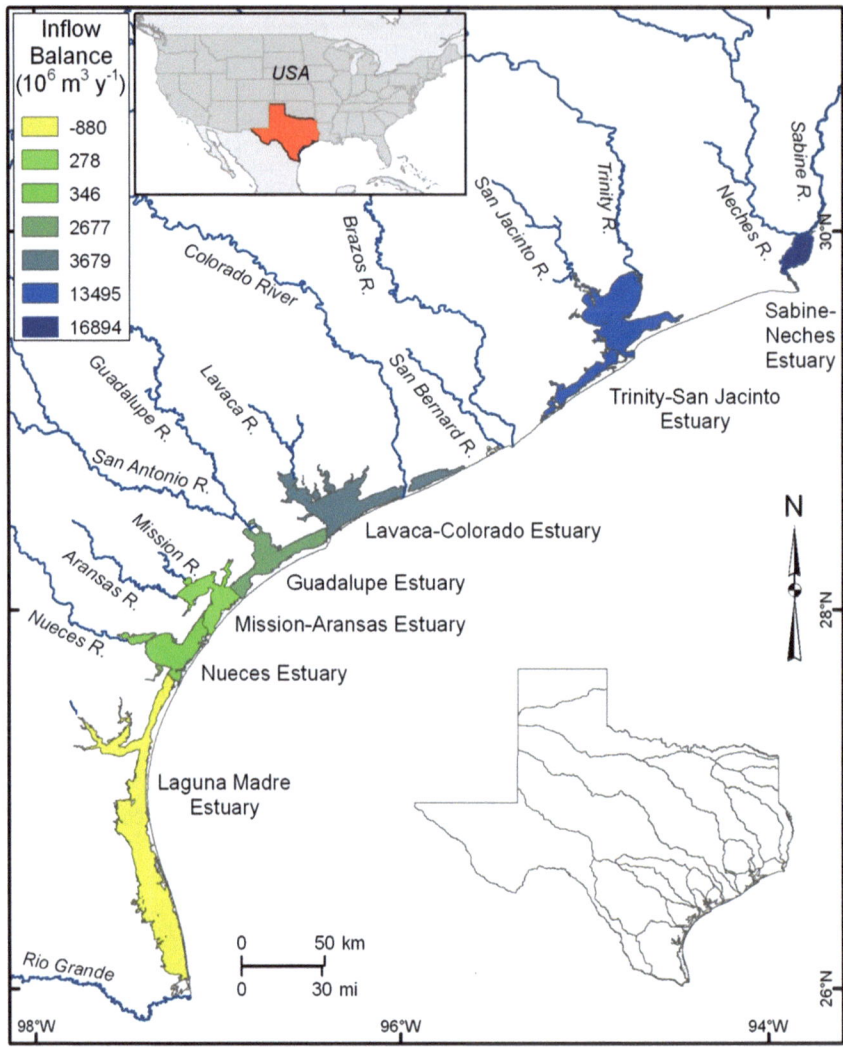

**Fig. 3.1** Location of Texas estuaries and inflow balance in each. Inflow data from the Texas Water Development Board. (http://midgewater.twdb.state.tx.us/bays_estuaries/hydrologypage.html)

draining watersheds (Fig. 3.1). The river generally flows into the secondary bay and thus secondary bays have greater freshwater influence. Primary bays provide the connection with the Gulf of Mexico and thus have greater marine influence.

The Texas coast is bounded by Sabine River (border to Louisiana) in the northeast and the Rio Grande (border with Mexico) in the southwest (Figs. 3.1 and 3.2). From northeast to southwest, the major bay-estuarine systems are the Sabine-Neches Estuary, Trinity-San Jacinto Estuary, Lavaca-Colorado Estuary, Guadalupe Estuary, Mission-Aransas Estuary, Nueces Estuary, and Laguna Madre Estuary.

Laguna Madre is actually two different systems: Upper Laguna Madre/Baffin Bay and Lower Laguna Madre. Texas follows the traditional system of naming an estuary for the river(s) that dilute sea water (Longely 1994). In NOAA publications (e.g., Orlando et al. 1993), these systems are named after the primary bay (Sabine Lake, Galveston Bay, Matagorda Bay, San Antonio Bay, Aransas Bay, Corpus Christi Bay, and Laguna Madre, respectively). There are also two riverine estuaries, the Brazos River and the Rio Grande, which flow directly to the Gulf of Mexico (Fig. 3.2).

The estuaries of Texas are remarkably hydrologically diverse in spite of similar geomorphology. This is due to a climatic gradient, which influences freshwater inflow to estuaries. The gradient of decreasing rainfall, and concomitant freshwater inflow, from northeast to southwest, is the most distinctive feature of the coastline (Fig. 3.3, Table 3.1). Along this gradient, rainfall decreases by a factor of two, but inflow balance decreases by almost two orders of magnitude (Fig. 3.1). Inflow balance is the sum of freshwater inputs (gaged, modeled runoff, direct precipitation, plus return flows) minus the outputs (diversions and evaporation). The net effect is a gradient with estuaries with similar physical characteristics but a declining salinity gradient.

Freshwater inflow balance patterns appear to group into four distinct climatic subregions, which vary by about an order of magnitude each The northeastern most subregion is composed of the Sabine-Neches Estuary (containing Sabine Lake) and the Trinity-San Jacinto Estuary (containing Galveston Bay). This northeastern subregion has the highest rainfall and inflow balance greater than $10^{10}$ $m^3$ $yr^{-1}$. The next three climatic subregions form the largest area, the Coastal Bend, which is

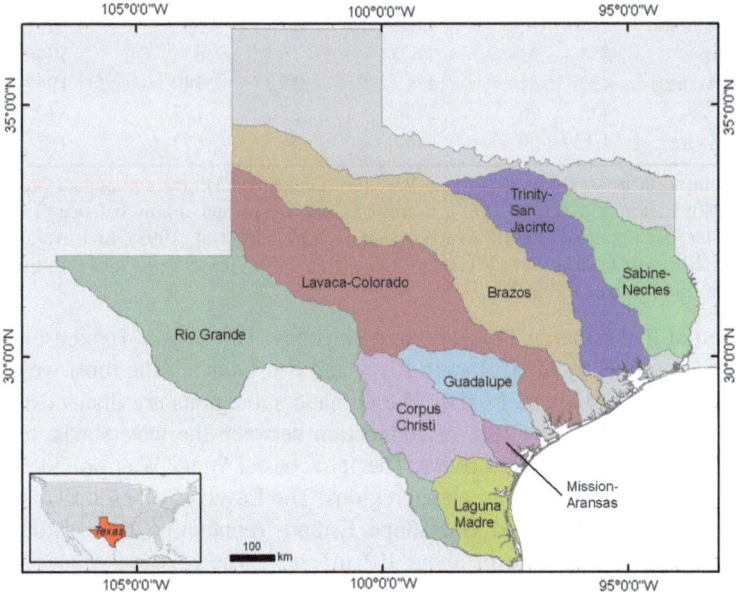

**Fig. 3.2** Catchments of the seven major Texas estuaries and two major River estuaries (Rio Grande and Brazos River). Only portions of the catchments that are in Texas are included in map

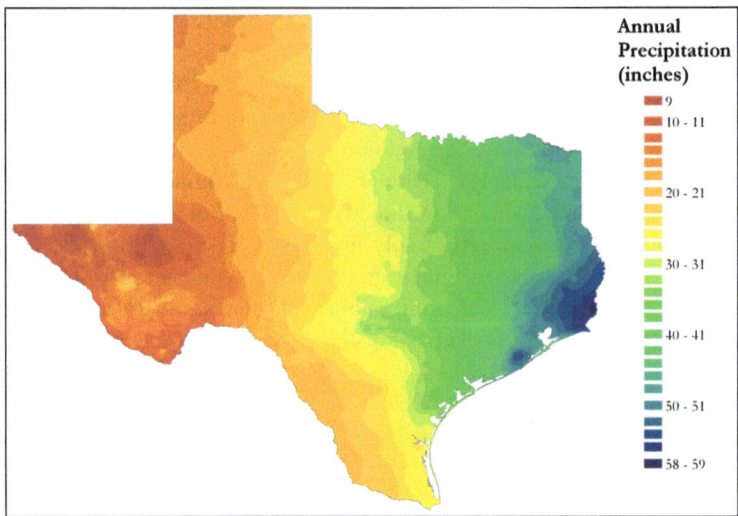

**Fig. 3.3** Precipitation gradient within Texas (data source: Texas General Land Office)

**Table 3.1** Climatic gradient in Texas Estuaries

| Estuary | Area (km$^2$) | Rainfall (cm yr$^{-1}$) | Inflow ($10^6$ m$^3$ yr$^{-1}$) | Salinity (ppt) | Commercial harvest Finfish ($10^3$ kg yr$^{-1}$) | Shellfish ($10^3$ kg yr$^{-1}$) |
|---|---|---|---|---|---|---|
| Sabine-Neches | 183 | 142 | 16,894 | 8 | 3 | 341 |
| Trinity-San Jacinto | 1,416 | 112 | 13,495 | 16 | 176 | 4352 |
| Lavaca-Colorado | 1,158 | 102 | 3,679 | 18 | 59 | 2531 |
| Guadalupe | 551 | 91 | 2,677 | 16 | 63 | 1846 |
| Mission-Aransas | 453 | 81 | 278 | 15 | 140 | 1947 |
| Nueces | 433 | 76 | 346 | 23 | 173 | 840 |
| Laguna Madre | 1,139 | 69 | −880 | 36 | 677 | 163 |

Listed from north to south: area at mean low tide (Diener 1975), average annual precipitation (1951–1980; Larkin and Bomar 1983), average annual freshwater inflow balance (1941–1994; Texas Water Development Board), average salinity (Orlando et al. 1993), and average annual commercial harvest (1962–1998; McEachron and Fuls 1996; Texas Parks and Wildlife Department 1988; Robinson et al. 2000)

composed of five estuaries linked by large lagoons. The Coastal Bend estuaries are in an area bounded by the Colorado River and Rio Grande. The most well-known lagoonal estuary is Laguna Madre. The climatic subregions are distinct in several ways. Most important is a lack of connection between the watersheds, thus each bay system is fed by different rivers. The Intracoastal Water Way provides a man-made, dredged channel linking all subregions. The Lavaca-Colorado Estuary (containing Matagorda Bay) and Guadalupe Estuary (containing San Antonio Bay) have an average inflow rate of about $10^9$ m$^3$ yr$^{-1}$. The Mission-Aransas Estuary (containing Aransas Bay) and Nueces Estuary (containing Corpus Christi Bay) have an average inflow rate of about $10^8$ m$^3$ yr$^{-1}$. Laguna Madre is a negative estuary because evaporation exceeds inputs and has an average negative inflow rate of about $10^8$ m$^3$ yr$^{-1}$. Thus, the region spans positive, neutral, and negative estuaries.

There is also a concomitant gradient of different timing of peak inflow events (Fig. 3.4). The northern estuaries receive peak inflow during the spring, the central estuaries are bimodal receiving peak inflows during the spring and fall, and the southern-most estuaries receive peak inflows during the fall. These distinct patterns are very important ecologically, because growth, reproduction, and migration of many species are keyed to seasonal events. The timing and magnitude of inundation is believed to regulate finfish and shellfish production (Texas Department of Water Resources 1982). The differences within and among the subregions and estuaries of Texas provides a sufficiently broad scale to examine effects of climate change and variability on ecological processes.

The latitudinal gradient of decreasing inflow into estuaries regulates salinity. As well as a latitudinal climatic gradient, there is a longitudinal salinity gradient within each estuary. The salinity gradient within and among the estuaries has already been demonstrated to regulate the infaunal molluscan community (Montagna and Kalke 1995). There are salinity gradients within the estuaries from the river mouth to the sea, which influences the zonation of communities found within the estuaries (Kalke and Montagna 1991; Montagna and Kalke 1992, 1995). The interactions among the geophysical factors of climate, estuarine physiography, and diversity of habitat types in the Gulf of Mexico are factors that influence diversity of the region.

Another characteristic of Texas estuaries is the extreme year-to-year variability in inflow (Fig. 3.5). Consequently, salinity gradients within estuaries vary from year to year. The southwestern estuaries in particular appear to be in a nearly

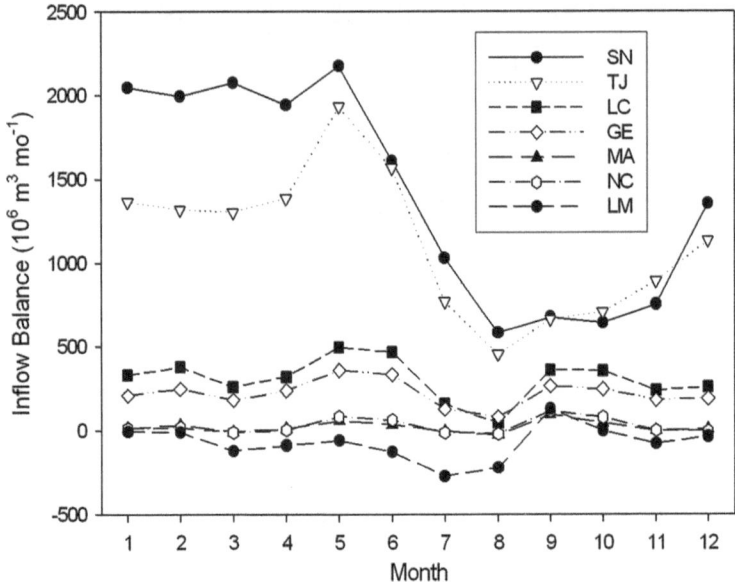

**Fig. 3.4** Average monthly inflow balance (1941–1994) in Texas estuaries. *SN* Sabine-Neches Estuary, *TJ* Trinity-San Jacinto Estuary, *LC* Lavaca-Colorado Estuary, *GE* Guadalupe Estuary, *MA* Mission-Aransas Estuary, *NC* Nueces Estuary, and *LM* Laguna Madre Estuary

desert climate that is punctuated by flood events. The floods are caused by tropical storms or larger global climate patterns. The El Niño Southern Oscillation (ENSO) has a strong influence on inflow to Texas estuaries. The Southern Oscillation Index (SOI; http://www.cgd.ucar.edu/cas/catalog/climind/soi.html) is negative during El Niño events. There is an inverse correlation between SOI and total inflow to the Texas coast ($r = -0.14$, $p = 0.0004$). The inverse correlation between smoothed

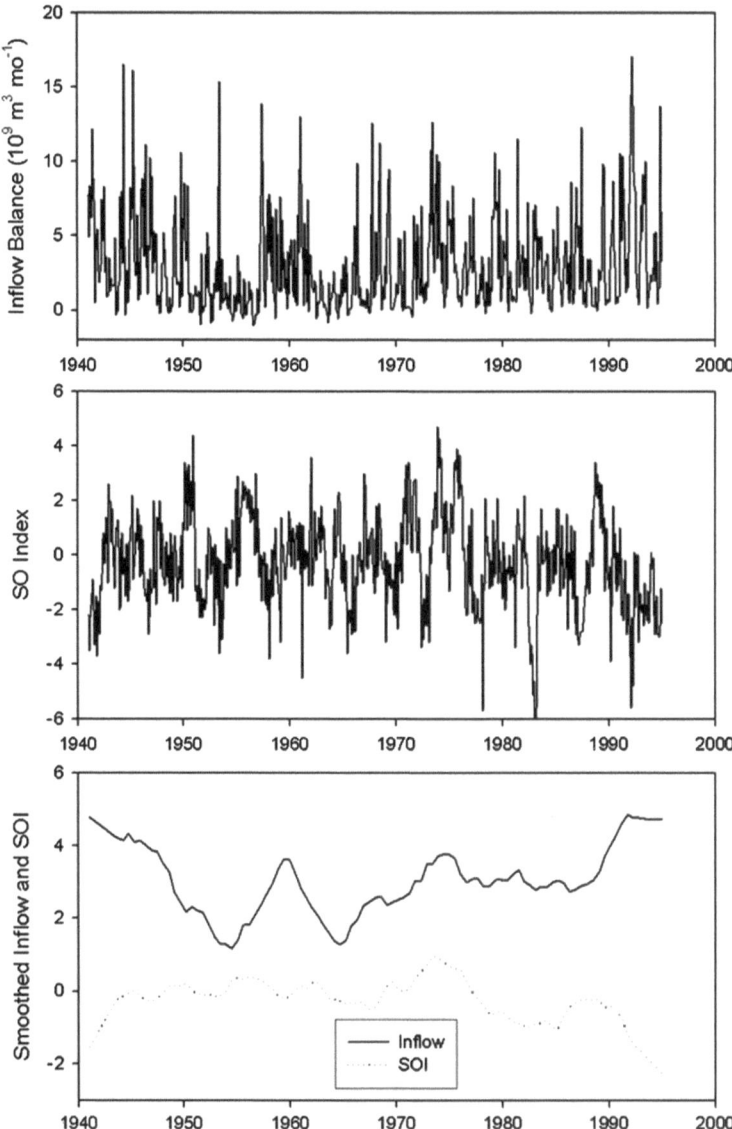

**Fig. 3.5**   Total inflow to Texas Coast and Southern Oscillation Index (SOI). *Bottom* panel contains trends smoothed using polynomial regression and weights computed from the Gaussian density

SOI and smoothed total inflow to the Texas coast is strong ($r = -0.47$, $p < 0.0001$; Fig. 3.5). The ENSO phenomenon is only one climatic factor affecting inflow. Inflow is also influenced by tropical waves, which affect the coast from the east.

The Texas coast is likely an ideal area to study climate change effects on estuaries, because it already is a natural experiment. There is physical similarity among Texas estuaries, each is simple draining only one or two watersheds, and they lie in a climatic gradient that is influenced by large-scale climate patterns in the Pacific and Atlantic Oceans. Being semiarid and semitropical, small changes in global temperature will likely have large effects. It is simple to posit hypotheses, e.g., drier conditions will result in estuaries more like the southwestern estuaries, and wetter conditions will result in estuaries more like those to the northeast. It will be possible to design stratified sampling programs where statistical control can be used on confounding factors, e.g., watershed drainage basins, anthropogenic inputs, Gulf of Mexico exchange, specific habitats, circulation patterns, and alterations by man.

## 3.1.2 History of Freshwater Management

The state of Texas first made efforts to examine the coast-wide problem of freshwater inflows following an almost 8-year drought in the 1950s that was so severe that 244 of 254 Texas counties were declared disaster areas (TWDB 2009) and many Texas rivers finally stopped flowing. The drought resulted in hypersaline estuaries, fish kills, loss of blue crabs and white shrimp, and invasions by stenohaline species (Copeland 1966; Hoese 1967). The most positive effect of the drought was the passing of the Texas Water Planning Act by Texas State Legislature in 1957. This act was amended over the next 10 years and led to the creation of a Texas Water Plan that was adopted in 1969. The plan called for an estimated 2.5 million acre-feet (3 billion $m^3$) of supplemental freshwater inflows annually to Texas bays and estuaries. In 1967, a cooperative Bays and Estuaries program was created to collect physical, chemical, and biological data to inform water planning. This program was expanded by legislative mandate in 1975 (Senate Bill 137), which required comprehensive studies of the effects of freshwater inflows on bays and estuaries. The results of these studies were published in a series of reports and summarized by the Texas Department of Water Resources (1982). Although the reports included preliminary estimates of freshwater inflow needs by each bay, it was generally accepted that the databases underlying those estimates were too limited to be used for water management (Longley 1994).

The need for further research on water management was supported by bills that were passed in 1985 (House Bill 2) and 1987 (Senate Bill 683) and as a consequence recommendations were developed to continue estuarine monitoring programs to provide better data to inform state water planning and permitting decisions (Longley 1994; Powell et al. 2002). In 1995, the state legislature extended the monitoring mandate and added new language to the Texas Water Code to "maintain those conditions considered necessary to maintain beneficial inflows to any affected bay or estuary system", where 'beneficial inflows' describe "a salinity, nutrient, and sediment loading regime adequate to maintain and ecologically sound environment"

(Texas Water Code 11.147 (a)(b)). Currently, state agency staff coordinates with water authorities, consultants, and university scientists to investigate each major estuary system using a standard protocol to develop freshwater inflow recommendations for implementation in state water planning and permitting processes. The Texas Parks and Wildlife Department, Coastal Fisheries Division, has an extensive monitoring program for fish in all Texas bays. The Texas Water Development Board monitors and collates river inflow and bay hydrographic data to estimate flows to the coast. These data are used in periodic assessments to revise inflow targets.

Texas legislators did not give a definition of an "ecologically sound environment", but many interpretations have been made. An estuary can be considered ecologically sound when the physical, biological, and chemical parameters that are measured fall within the range of values that existed before human interference (Longley 1994). Ecological soundness can also be characterized as an environment that has several trophic levels through which nutrients are cycled, so that production of commercially and recreationally important fisheries is maintained. With these goals in mind, Texas state agencies have been working to investigate the relationships between freshwater inflow and salinity, nutrient and sediment loadings, to determine the conditions required for and ecologically sound environment, and to develop methods to quantify freshwater inflow needs (Longley 1994).

As described in Sect. 2.2, there is a way to define ecological soundness based on the concepts the US Environmental Protection Agency uses for defining assessment of ecological health in the water quality context. Ecological health is assessed by determining if *indicators* of ecological *conditions* are in an acceptable range. *Indicators* are measures (or metrics) of ecological health for which sufficient information exists to establish an acceptable range of responses across broad spatial and temporal scales. For example, in our own work, ecological health or soundness has been defined by benthic metrics. The condition of integrity is assessed when community structure and diversity are stable over long time periods. The condition of function is assessed when biomass (the best indicator of productivity) is stable over long time periods. These three metrics: abundance, biomass, and species are easily obtainable in routine sampling programs.

There was great concern that the existing policies were not protective of environmental flows, so a study was commissioned by the Texas Legislature to assess environmental flows state-wide (SAC 2004). When no action was taken in the 2005 Legislative session, the Governor requested an update to the study (SAC 2006). In 2007, the Texas Legislature made a substantial change to water policy with the passage of Senate Bill 3 (SB3). Senate Bill 3 requires that any new water permits contain, to the extent possible, a set aside for an environmental flow regime. The new regulatory approach to protect water courses requires environmental flow standards to be developed by the Texas Commission on Environmental Quality (TCEQ) rulemaking process. Senate Bill 3 directed the use of an environmental flow regime in developing flow standards and defined an environmental flow regime as a schedule of flow quantities that reflects seasonal and yearly fluctuations that typically would vary geographically, by specific location in a watershed, and that are shown to be adequate to support a sound ecological environment and to maintain the productivity, extent, and persistence of key aquatic habitats.

As a part of SB3, the Environmental Flows Advisory Group (EFAG) was created 'in recognition of the importance that the ecological soundness of our riverine, bay, and estuary systems and riparian lands has on the economy, health, and well-being of the state' (Texas Water Code, Section 11.0236). The EFAG is composed of policy makers: three Texas House members, three Texas Senate members, and the three heads of the natural resource agencies. A Science Advisory Committee (SAC) was created to advise the EFAG (Fig. 3.6). The main role of the SAC is in recommending technical guidance to assist in a consistent approach state-wide. To fulfill this role, the SAC has created nine guidance documents (SAC 2009a, b, c, d, e, f, 2010a, b, and c).

While SB3 did not define a "sound ecological environment," the SAC (2006) suggested that a "sound ecological environment" is one that: 1) sustains the full complement of native species in perpetuity, 2) sustains key habitat features required by these species, 3) retains key features of the natural flow regime required by these species to complete their life cycles, and 4) sustains key ecosystem processes and services, such as elemental cycling and the productivity of important plant and animal populations. This definition was also offered as guidance for implementation by the SB3 SAC, which further noted that "underlying each of these recommendations is the need to establish relationships between elements of the environment, including flows, and the native species and their functions" (SAC 2009c).

SB3 created an adaptive management process whereby scientific and stakeholder groups are formed for individual estuary/basin systems called Basin and Bay Expert Science Team (BBEST), and Basin and Bay Area Stakeholder Committee (BBASC), respectively. These estuary-focused groups interact with each other to inform the Texas state agencies, and ultimately the Texas Commission on Environmental Quality, of their recommendations for the quantity, quality, and sources of freshwater inflow that reach each estuary (Fig. 3.6).

The BBEST teams from all the basins have submitted Final Environmental Flow Recommendations reports and these can be downloaded from the Environmental Flows Resources page maintained by the TCEQ (http://www.tceq.

**Fig. 3.6** The Texas environmental flows allocation process overview. *Arrow* direction indicates flow of information (adapted from Hess 2010)

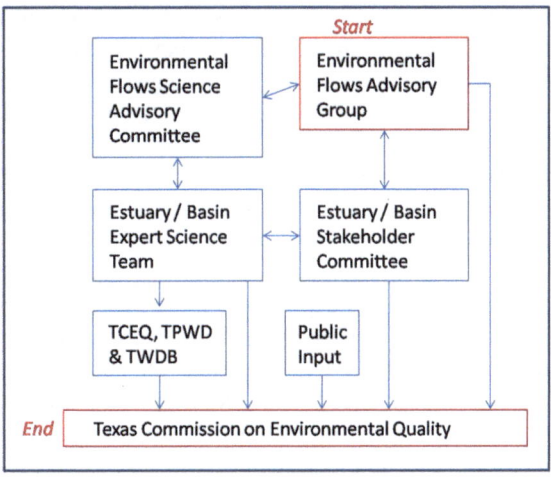

state.tx.us/permitting/water_supply/water_rights/eflows/resources.html).        This website also contains all the documentation generated during the SB3 process, which includes the SAC guidance, BBEST reports, BBASC reports, and TCEQ rules.

While most observers expected the different science teams to use different approaches, there was remarkable consensus among the teams in methodology. All used a salinity zone approach to recommend inflow regimes. Bioindicators were chosen for various salinity zones, and the salinity requirements of those indicators were identified. Then, flows were identified to yield the desired salinity conditions during various periods and with various frequencies.

One specific case for example is the Trinity and San Jacinto Rivers and Galveston Bay BBEST (TJBBEST). This group actually offered two different recommendations: One recommendation was to adopt a single annual value with no regimes until more data could be obtained (which is referred to as the "conditional" recommendation. The second opinion (which is referred to as the "regime group" recommendation) used salinity as a proxy for freshwater inflow and therefore provides suitable salinity criteria rather than inflow volumes for the estuary (TJBBEST 2009). Salinity criteria were identified for several biological indicator species (Table 3.2) in the regions that they currently occur (Fig. 3.7). The TJBBEST also identified applicable time periods to implement salinity criteria for each species (period of concern) rather than to apply salinity criteria permanently. The spatial and temporal salinity ranges of the biological indicator species were determined from long-term monitoring databases from Texas Parks and Wildlife Department (a state resource agency) or other monitoring studies. Most biological indicator species selected were sessile organisms, as these organisms integrate changes in water quality over time while remaining in the same place.

The Sabine and Neches Rivers and Sabine Lake BBEST (SNBBEST) used an instream flow focus to determine optimal inflow to the Sabine-Neches Estuary (Hess 2010). The Sabine-Neches Estuary, located on the Texas-Louisiana border, is unique among Texas Estuaries in that it is the smallest major estuary in Texas, receives the most freshwater inflow, and has a median salinity of only 8 psu (Longely 1994; SNBBEST 2009). The Sabine-Neches estuary is also unique because its salinity regime has changed more from deepening of the ship channel than because of a reduction in freshwater inflows. The SNBBEST created subcommittees for research on gaging, hydrology, biology, water quality, and geomorphology. SNBBEST recommended implementing seasonal subsistence flows as the minimum amount of inflow to the estuary. Seasonal subsistence flows were calculated using existing gage flow data as being equivalent to the median of the lowest 10 % of historical base flows by season. The SNBBEST also recommended seasonal base flows depending on the hydrologic condition of the season depending on if it was wet, average, or dry. Seasonal pulse flows into the estuary are also recommended but at a rate less than the current one to two events per season. The inflow approach used by the SNBBEST also stresses revisions of the current criteria in the future as more knowledge is attained and consequences of such recommendations are implemented.

**Table 3.2** Identified biological indicators for evaluating freshwater inflow needs to Galveston Bay (from TJBBEST 2009)

| Habitat indicator | Common name | Scientific name | Criterion | Period of concern |
|---|---|---|---|---|
| | Wild Celery | *Vallisneria americana* | <5 psu for germination and establishment | Spring |
| | Wild Celery | *Vallisneria americana* | <10 psu for survival | Summer and Fall |
| Low-salinity indicators | Atlantic Rangia | *Rangia cuneata* | 2–10 psu for spawning and larval survival | Spring and Fall |
| | Gulf menhaden | *Brevoortia patronus* | 5–15 psu for occurrence as forage fish | Winter and Spring |
| | Blue catfish | *Ictalurus furcatus* | <10 psu for occurrence as predator | Single pulse in winter or spring |
| High salinity indicators | Mantis shrimp | *Squilla empusa* | >25 psu for abundance | Summer–Fall |
| | Pinfish | *Lagodon rhomboides* | >25 psu for abundance | Summer–Fall |
| Oyster health indicators | Dermo and oyster drill impacts on oyster | Dermo = *Perkinsus marinus* Oyster drill = *Stramonita haemastoma* | 10–20 psu to prevent excessive parasitism and predation | July–September |
| | | Oyster = *Crassostrea virginica* | <5 psu to remove parasite load from central reefs | 2 weeks at 10-year intervals |

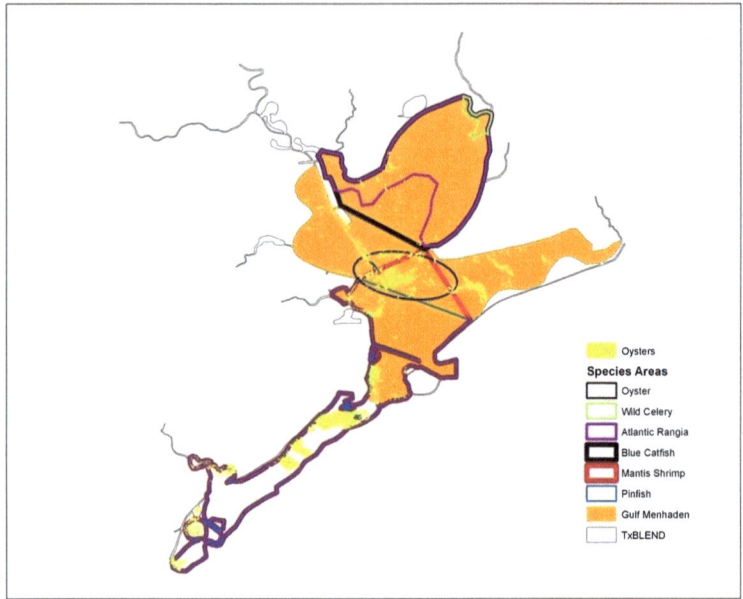

**Fig. 3.7** Map of the areas assigned to each biological indicator for evaluation of flow effects on the suitability of the salinity conditions (from TJBBEST 2009)

### 3.1.3  Recent Texas Research

An important objective of many resource agencies is to quantify the relationship between bioindicators of marine resource populations and freshwater inflows to bays and estuaries. However, there is year-to-year variability in population densities and successional events in estuarine communities. This year-to-year variability is apparently driven by long-term and global-scale climatic events, e.g., El Niño, which affects rates of precipitation and concomitantly rates of freshwater inflow (Tolan 2007). Therefore, the best approach is to document long-term changes in populations and communities that are influenced by freshwater inflow. The best indicator of productivity is the change in biomass of the community over time (Banse and Moser 1980). Based on initial sampling of 1–4 years of benthic data in Texas bays, it was originally concluded that inflow does increase benthic productivity (Kalke and Montagna 1991; Montagna and Kalke 1992, 1995). However, further analysis of the data set over a 5-year period demonstrated that the largest effect may not be on productivity, but may be on community structure (Montagna and Li 2010). This implies that reduced inflows may not only reduce productivity, but may also change the composition of species in an estuary.

Texas Coastal Bend estuaries were studied over a 20-year period by to determine the long-term response of benthic organisms to freshwater inflow (Montagna 2008; Montagna and Kim 2012). Results show that the biological effects on benthic communities appear to be driven by the El Niño cycle. Flood conditions introduce nutrient rich waters into the estuary that result in lower salinity. During El Niño periods, the lowest salinities and highest nutrient values were recorded. During these periods, the spatial extent of the freshwater fauna is increased, and the estuarine fauna replaced the marine fauna in the lower end of the estuary. The high level of nutrients stimulated a burst of benthic productivity (of predominantly freshwater and estuarine organisms), which lasted about 6 months. This was followed by a transition to a drought period with low inflow resulting in higher salinities, lower nutrients, dominance by marine fauna, decreased productivity and abundances.

The Lavaca and Matagorda Bay system on the central Texas coast was studied to investigate the relationship between benthic macrofaunal community structure and sediment and hydrologic parameters (Pollack et al. 2009). A principal component analysis (PCA) was used to illustrate the relationships among water column environmental variables (Fig. 3.8). The PC 1 variable loads have high negative values for silicate ($SiO_4$), dissolved inorganic nitrogen (DIN; sum of nitrate, nitrite and ammonium), and phosphate ($PO_4$); and high positive values for salinity. High salinity is correlated with low nutrients (and vice versa), thus PC 1 represents a linear scale of freshwater inflow effects. High temperature is correlated to low dissolved oxygen (DO) along PC 2, thus PC 2 represents seasonal effects. PC 1 (freshwater inflow effects on hydrology) was significantly correlated with

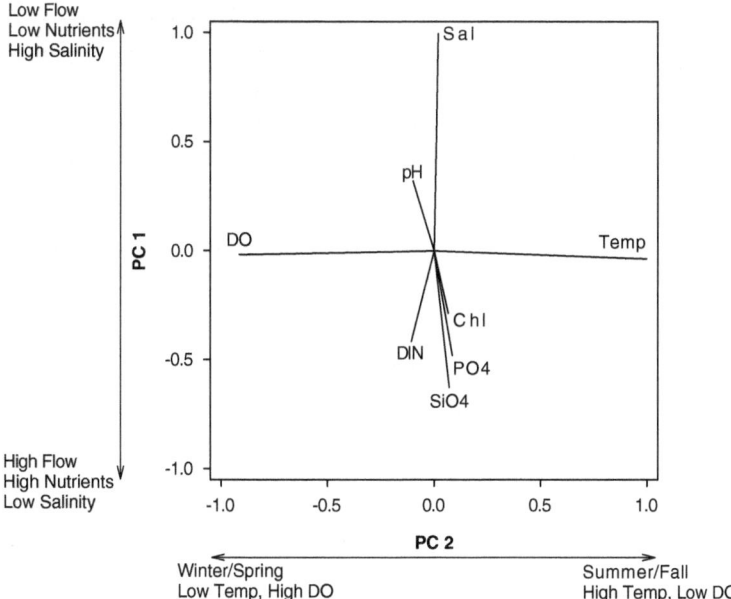

**Fig. 3.8** Principal component analysis (PCA) showing relationships between environmental variables in Lavaca-Colorado Estuary, Texas (from Pollack et al. 2009). Principal components 1 and 2 (PC 1 and PC 2) explained 47 and 20 % of the variance. *Sal* salinity, *DIN* dissolved inorganic nitrogen (sum of nitrate, nitrite and ammonium), *PO₄* phosphate, *SiO₄* silicate, *Chl* chlorophyll, *DO* dissolved oxygen, *Temp* temperature

macrobenthic diversity and evenness, while sediment characteristics did not have a significant relationship to any biological variables (Table 3.3).

In a related study of the Matagorda Bay system, a bioenergetic model was created to predict the benthic biomass effects that could result from changes in freshwater inflow (Kim and Montagna 2009). Model simulation results were interpreted to assess the role of freshwater inflow in controlling benthic productivity. In general, when salinity increased with decreasing nutrient concentrations, deposit feeder biomass increased while suspension feeder biomass decreased (Fig. 3.9). The change in trophic groups also indicates that functional diversity will decrease with increasing salinity. In other words, reduced inflow will change the character of the bay into one that produces mainly worms, and not worms and shellfish. Estuary-wide, the model predicts that reduced freshwater inflow could increase benthic productivity overall as upstream communities acclimate to elevated salinities and take on characteristics of downstream polyhaline communities.

The 50-month Rincon Bayou Demonstration Project is an example of a recent project investigating freshwater inflow needs of bays and estuaries in the state of Texas. The demonstration project increased the amount of freshwater into the

**Table 3.3** Linear correlations between sediment and hydrologic principal component factor scores and biological variables in the Lavaca-Colorado Estuary from Pollack et al. (2009)

| Benthic metric | Sediment | | Water column | |
|---|---|---|---|---|
| | PC1 | PC2 | PC1 | PC2 |
| Biomass (g m$^{-2}$) | −0.166 | 0.406 | −0.164 | −0.305 |
| | 0.510 | 0.094 | 0.075 | 0.001 |
| | 18 | 18 | 119 | 119 |
| Abundance (n m$^{-2}$) | −0.290 | 0.311 | −0.055 | 0.059 |
| | 0.242 | 0.209 | 0.556 | 0.525 |
| | 18 | 18 | 119 | 119 |
| Diversity (N1) | 0.055 | 0.445 | −0.289 | −0.407 |
| | 0.829 | 0.064 | 0.001 | <0.0001 |
| | 18 | 18 | 119 | 119 |
| Diversity (H′) | 0.159 | 0.452 | −0.310 | −0.408 |
| | 0.528 | 0.060 | 0.001 | <.0001 |
| | 18 | 18 | 119 | 119 |
| Evenness (J′) | 0.193 | 0.291 | −0.208 | 0.122 |
| | 0.442 | 0.242 | 0.023 | 0.187 |
| | 18 | 18 | 119 | 119 |
| FIBI PC1 | 0.320 | 0.179 | −0.198 | −0.068 |
| | 0.196 | 0.478 | 0.033 | 0.464 |
| | 18 | 18 | 117 | 117 |
| FIBI PC2 | −0.302 | 0.386 | −0.150 | −0.515 |
| | 0.222 | 0.114 | 0.106 | <0.0001 |
| | 18 | 18 | 117 | 117 |

$r$ Pearson product correlation coefficient, $P$ probability of the null hypothesis, $n$ number of sample pairs

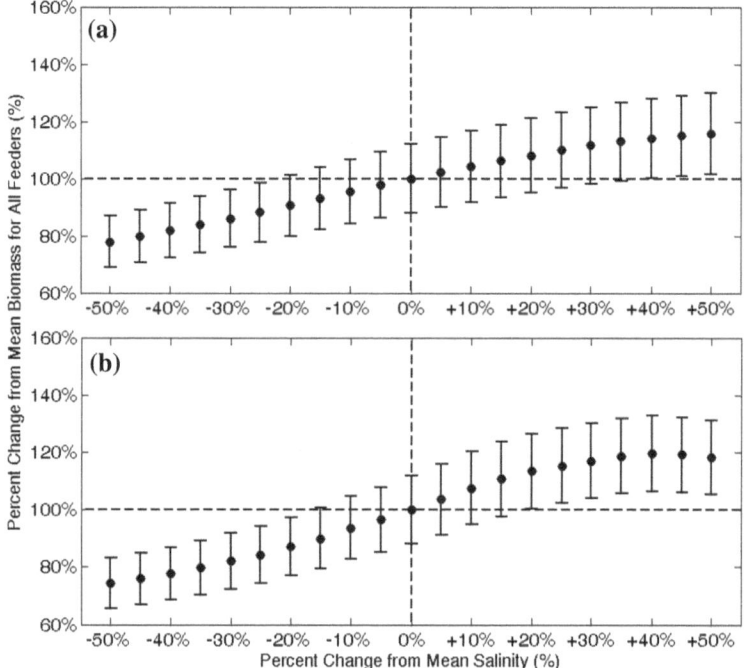

**Fig. 3.9** Predicted percent change from mean biomass for both deposit and suspension feeders (*solid circles*) based on changes (–50 to 50 %) of long-term mean salinity and nutrients in **a** Lavaca Bay and **b** Matagorda Bay (from Kim and Montagna 2008). The *error bars* represent the standard error. The number of simulated data points during 1988–2005 was 69 data points for Lavaca and 72 for Matagorda Bay. The long-term mean salinity is 15.1 for Lavaca Bay and 24.1 for Matagorda Bay, and these values represent long-term mean salinity with 0 % change. The long-term mean biomass for all feeders was 2.36 g dw m$^{-2}$ for Lavaca Bay, and 12.78 g dw m$^{-2}$ for Matagorda Bay, respectively, and these values represent long-term mean biomass under 0 % of salinity. Labels the X-axis represent salinity in percentage and corresponding changes in nutrients are not shown here but were considered in the model simulations. *Dashed lines* represent the position of reference simulation (0 % change)

upper Nueces Delta by about 732 %, reversing the hypersaline salinity gradient (>50 psu) to a more natural pattern (21–28 psu, Bureau of Reclamation 2000). Benthic organisms were better able to tolerate brackish salinity conditions than hypersaline conditions. As a result, total macrofauna abundance and size of individuals increased, particularly within the salinity range of 10–45 psu. Macrofauna characteristics demonstrated strong nonlinear relationships with salinity (Fig. 3.10; Montagna et al. 2002b). Macrofauna abundance appeared to peak at a salinity of 33 psu, biomass peaked at a salinity of 19 psu, and diversity peaked at a salinity of 9 psu. As a secondary result, the abundant and diverse benthic community of the demonstration project provided conditions favorable for feeding and growth of fish and shrimp. Fish and shrimp utilize estuaries as seasonal nursery areas (Riera et al. 2000). A novel log-normal regression model is used to identify

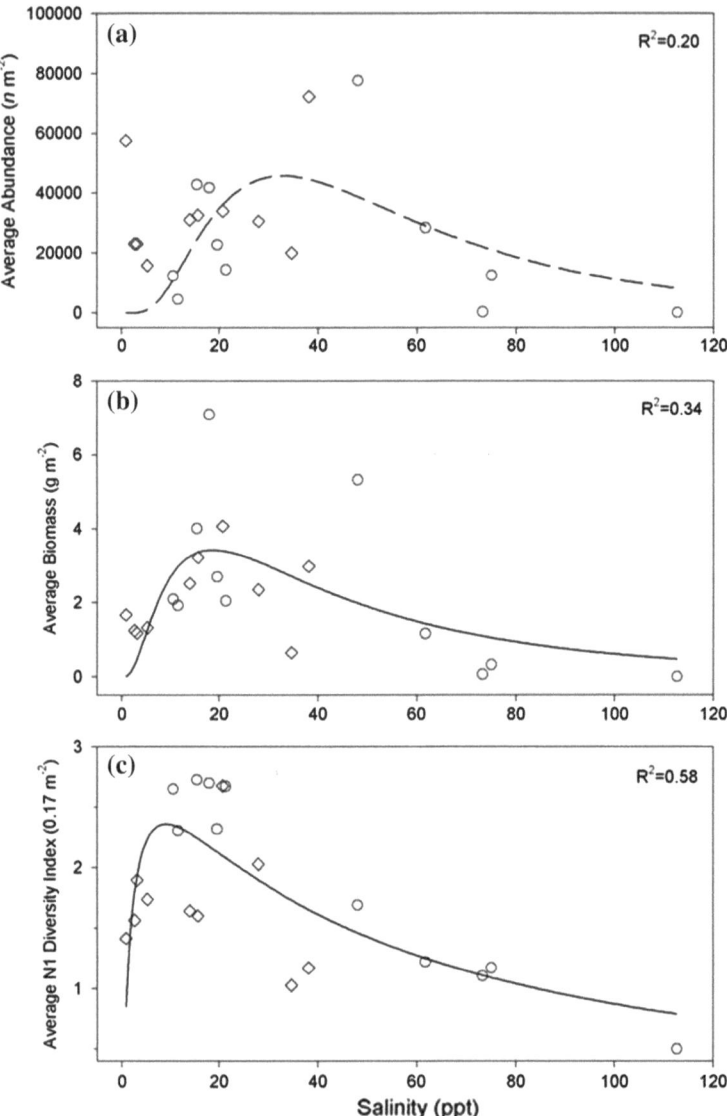

**Fig. 3.10** Macrofauna community response as a function of salinity in Rincon Bayou, Texas (from Montagna et al. 2002b). Abundance (**a**), biomass (**b**), and N1 diversity (**c**). Nonlinear response to salinity (*solid line*) at each time period. *Circles* represent periods of rising salinity and *diamonds* represent periods of falling salinity. *Dashed line* is not significant

the salinity range and inflow range that yield the peak abundance, biomass, and diversity (Montagna et al. 2002b). When macrofaunal characteristics were regressed against nonzero cumulative quarterly inflow rates, only biomass was significantly correlated, with a peak macrofauna biomass of 4 g m$^{-3}$ occurring

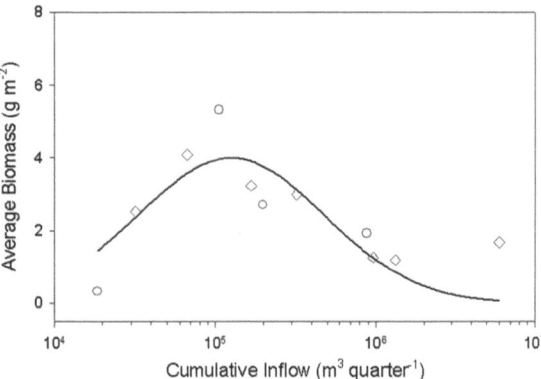

**Fig. 3.11** Cumulative inflow versus macrofaunal biomass in Rincon Bayou, Texas from Montagna et al. (2002b). *Circles* represent periods of rising salinity and *diamonds* represent periods of falling salinity

at flow rates of 125,000 m$^3$ per quarter (Fig. 3.11). So, in this ecosystem, a little water went a long way to increase ecosystem health. Increased freshwater inflow was found to also affect the processes of nutrient cycling, increasing water column and marsh plant production.

At this point of time, there is no question as to whether freshwater inflow is important to coastal ecosystems. Rather, the important questions are how, when, where, and in what quantities inflow should be utilized for environmental purposes (JCSCWEF 2004). Scientific methods and protocols have advanced over the past 40 years to improve our understanding of the importance of freshwater inflow. One of the main conclusions of this work is that adaptive management and precautionary principle methods need to be incorporated into the scientific study, management strategy implementation, and regulatory permitting phases of future freshwater inflow studies (JCSCWEF 2004). This important conclusion should be applied coast-wide (throughout the USA) to develop future approaches of managing freshwater inflow for adequate protection of coastal ecosystems.

## 3.2 Florida, USA

Florida has a long history of water flow management, particularly in the southern portion of the state. In 1972, the Florida Water Resource Act was established, which created five Water Management Districts throughout the state (1972 Laws of Florida Ch. 72–299): (1) Northwest Florida Water Management District, (2) Suwannee River Water Management District, (3) St. John's River Water Management District, (4) Southwest Florida Water Management District, and (5) South Florida Water Management District (Fig. 3.12).

Florida defines the need for inflows broadly as "the limit at which further withdrawals would be significantly harmful to the water resources or ecology of the area." To allow successful planning and management of regional water resources, the districts were created with boundaries defined by watersheds rather than

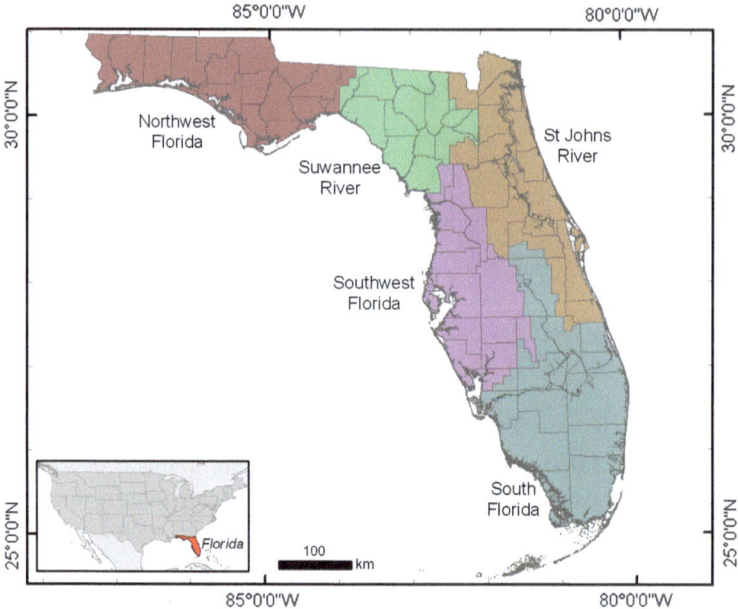

**Fig. 3.12**  Location of Water Management Districts, Florida, USA

political margins. Each district develops a water management plan for water resources within its region that identifies their role in water resource management and provides long-term (20-year) direction. The plans are updated at least every 5 years. The result is that freshwater inflow studies and rules are unique to each estuary and the different Florida water management districts have adopted different strategies for regulating freshwater flows in response to different intensities of human engineering and societal expectations of water management (e.g. flood control versus water supply).

## 3.2.1  Water Management District Profiles

The Northwest Florida Water Management District comprises 29,280 km$^2$ in the Florida panhandle, encompassing 16 counties and 6 major drainage basins (NWFWMD 2010). The district's population in 2007 was 1.3 million residents, or approximately 7.5 % of Florida's population.

The Suwannee River Water Management District is the smallest in Florida, encompassing 19,788 km$^2$ in the north-central region, and representing 15 counties and 13 river basins (SRWMD 2010). The total population of the district is expected to increase from 320,000 in 2010 to 750,000 by 2050 (SRWMD Strategic Plan 2010).

The St. John's River Water Management District comprises 32,116 km$^2$ in northeast and east-central Florida, including 18 counties and 10 major drainage basins (SJRWMD 2010). The St. John's River is the longest river in Florida, measuring 499 km and flowing south to north with a downhill gradient of less than 5 cm per mile (1.6 km) from head to mouth. From 1995 to 2000, the population of the St. John's River Water Management District increased from approximately 3.5–3.9 million people, an increase of about 11 % (SJRWMD DWMP 2005).

The Southwest Florida Water Management District is 25,900 km$^2$, encompassing 16 counties and 11 major drainage basins in west-central Florida (SWFWMD 2010). In 2000, the district's population was approximately 4 million people (SWFWMD DWMP 2005).

The South Florida Water Management District covers 16 counties and 2 major drainage basins in the southern half of the state (SFWMD 2010). The district includes Lake Okeechobee, the Everglades Agricultural Area, and the Florida Keys. The South Florida Water Management District's population is approximately 7.5 million people.

## 3.2.2 Florida Water Resource Policy

Chapter 373 of the Florida Statutes identifies four areas of responsibility for the districts: water supply, water quality, flood protection and floodplain management, and natural systems. The Water Management Districts were given a mandate to establish minimum flows and levels for surface waters and aquifers. Minimum flow was defined as "the limit at which further withdrawals would be significantly harmful to the water resources of ecology of the area."

According to Florida Administrative Code (FAC), minimum flows and levels should be evaluated to ensure consideration of natural seasonal fluctuations in water flows or levels, nonconsumptive uses, and environmental values associated with coastal, estuarine, riverine, spring, aquatic, and wetlands ecology, including the following 10 natural resource and environmental values: (1) recreation in and on the water, (2) fish and wildlife habitats and the passage of fish, (3) estuarine resources, (4) transfer of detrital material, (5) maintenance of freshwater storage and supply, (6) esthetic and scenic attributes, (7) filtration and absorption of nutrients and other pollutants, (8) sediment loads, (9) water quality, and (10) navigation (Section 62-40.473 F.A.C.).

In 1995, the Florida legislature concluded that none of the water management districts had fully completed the task of establishing minimum flows and levels (OPPAGA 1995), and the Florida Water Resource Act was amended in 1997 to require the development of a priority list for the establishment of minimum flows. This legislation is significant because (1) it states that resource-based criteria are central to the determination of minimum flows, (2) the objective of the minimum flow must be to sustain the resource, and (3) the determination of minimum flows is a bottom-up process with scientists determining the criteria and definitions of sustainability (Estevez 2002). The management districts have taken a variety of approaches in complying with this mandate (Alber 2002).

### 3.2.3  District Management Approaches

The Southwest Florida Water Management District has been utilizing a management approach for unimpounded rivers that limits withdrawals to a percentage of streamflow at the time of withdrawal (Flannery et al. 2002). The natural flow regime of a river is the baseline for identifying the effects of increased withdrawal levels; various streamflow parameters are then evaluated to determine changes in river flow regimes. This approach to water supply planning and regulation is designed to maintain the physical structure and ecological characteristics of unimpounded rivers. Relationships between freshwater inflow and estuarine characteristics are then examined to determine withdrawal limits that will not result in negative environmental impacts. This percent-of-flow approach was supported by initial findings that showed a curvilinear response of isohaline locations to freshwater inflow and the influence of inflow on catch per unit effort for a number of key organisms (Flannery et al. 2002). Studies of the Alafia River estuary in particular demonstrated that the abundances of many estuarine-resident and juvenile estuarine-dependent species decline during low-flow periods (Flannery et al. 2002). The Southwest Florida Water Management District has used the "percent-of-flow" approach to inform adaptive management, such that continued data collection is being used to improve management strategies over time.

In contrast, the South Florida Water Management District and the Suwannee River Water Management District use resource-based approaches for setting inflow requirements with the aim of protecting biodiversity, productivity, and connections to adjacent floodplains and downstream habitats (Alber 2002; Mattson 2002). In this approach, critical habitats (e.g., oyster reefs, vegetated communities, or tidal creeks), that are also referred to as valued ecosystem components (VEC), are identified as are the range of salinities experienced in these areas. The freshwater inflow regime needed to establish these salinity regimes is then determined and correlations between salinity and biological variables are examined (Mattson 2002). This management approach is tied to the concept that growth and survival of larval and juvenile fish and shellfish can be enhanced if favorable salinity and suitable physical habitat overlap spatially at the appropriate time of year (Browder and Moore 1981). Results from oyster monitoring in the Suwannee River Estuary showed that oyster reef development is best in areas influenced by freshwater inflow, with salinity requirements and timing of inflow being important for reproduction and successful spat settlement (Mattson 2002). In the Caloosahatchee Estuary, the proposed minimum inflow rule is based on protecting the habitat for three seagrass species that are sensitive indicators of inflow effects (Doering et al. 2002). The most vexing problem in Florida is the Everglades, because sheet flow to this icon of the state has been reduced dramatically over the last few decades. Planning is underway for a large replumbing of south Florida to resolve this problem.

The St. John's River Water Management District evaluated topographic, soil, and vegetation data collected within the plant communities associated with the river in conjunction with hydrologic modeling to determine minimum flows and levels (ECT 2008). Minimum flows and levels were evaluated for five criteria: inundation of riparian wetlands for stream biota, saturation of hydric hammocks, maintenance of riparian hydric soils, adequate depths for fish passage and eelgrass beds, and protection of eelgrass beds from boat and canoe traffic (CH2M Hill 1999). Salinity targets were developed to protect various biological resources including hard clams *Mercenaria mercenaria*, eastern oysters *Crassostrea virginica*, fish species red drum *Sciaenops ocellatus*, snook *Centropomus undecimalis*, and spotted seatrout *Cynoscion nebulosus*, and seagrasses such as *Halodule* and *Syringodium* (Estevez and Marshall 1993).

The Northwest Florida Water Management District has not yet established minimum flows and levels for its drainage basins. Seven water bodies have been identified as priority areas for establishing minimum flows and levels: Floridian Aquifer, Inland Sand and Gravel Aquifer, Deer Point Lake, Wakulla Spring, Jackson Blue Spring, Yellow River, and Morrison Spring (NWFWMD 2010). Criteria for identification of the existence or potential for significant harm to these water bodies included: (1) potential migration of saline water due to aquifer drawdown, (2) potential future water supply, (3) reduced discharge to coastal areas, (4) aquatic resources needs, and (5) consumptive demands.

### 3.2.4  Southwest Florida Mollusks

Six tidal rivers in southwest Florida along the Gulf of Mexico were examined using a meta-analysis of existing data sets of mollusk communities to determine salinity-mollusk relationships at regional scales (Fig. 3.13) (Montagna et al. 2008a).

The most important variable correlated with mollusk communities was salinity, which is a proxy for freshwater inflow. Mollusk abundance and diversity relationships with salinity were examined with a nonlinear model. The assumption behind the model is that there is an optimal range for salinity and values decline prior to and after meeting this maximum value. That is, the relationship resembles a bell-shaped curve. The shape of this curve can be predicted with a three-parameter, log-normal model:

$$Y = ae^{\left[-0.5\left(\frac{\ln\left(\frac{X}{X_c}\right)}{b}\right)^2\right]}$$

The model was used to characterize the nonlinear relationship between a biological characteristic ($Y$) and salinity ($X$). The three parameters characterize different attributes of the curve, where $a$ is the maximum value, $b$ is the skewness or rate of change of the response as a function of salinity, and $c$ is the location of the peak response value on the salinity axis.

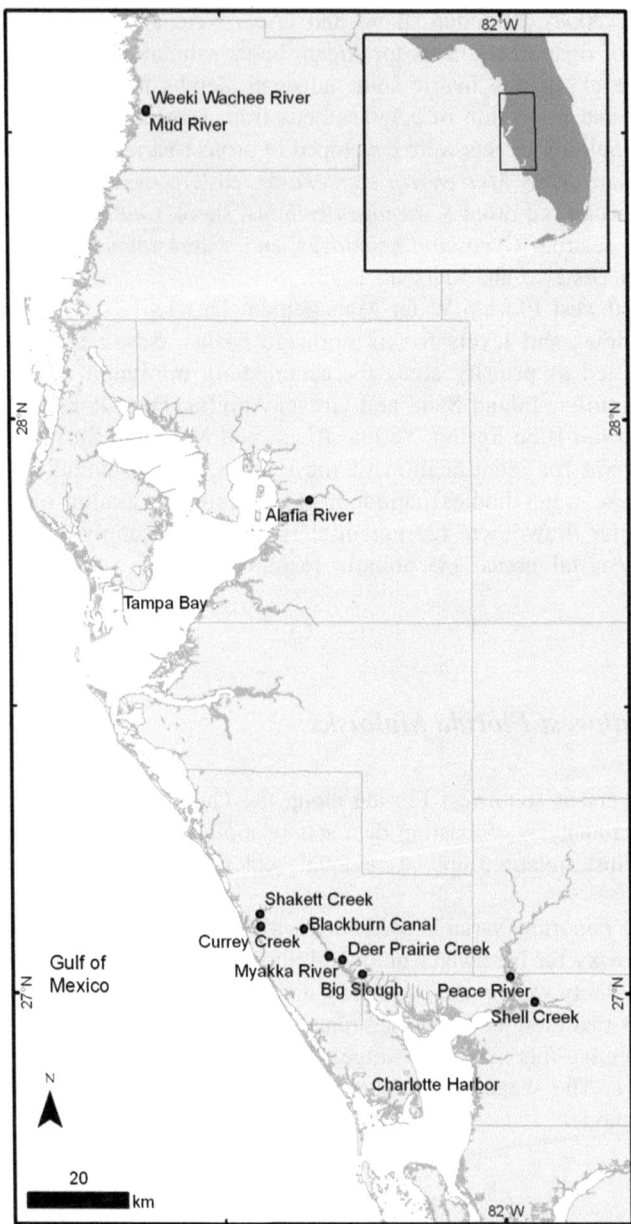

**Fig. 3.13** Map of the west coast of Florida showing the study sites

In the Peace and Myakka Rivers, there were strong relationships between mollusk diversity and abundance with salinity, where diversity and abundance increased with increasing salinity, peaked, and then declined (Fig. 3.14).

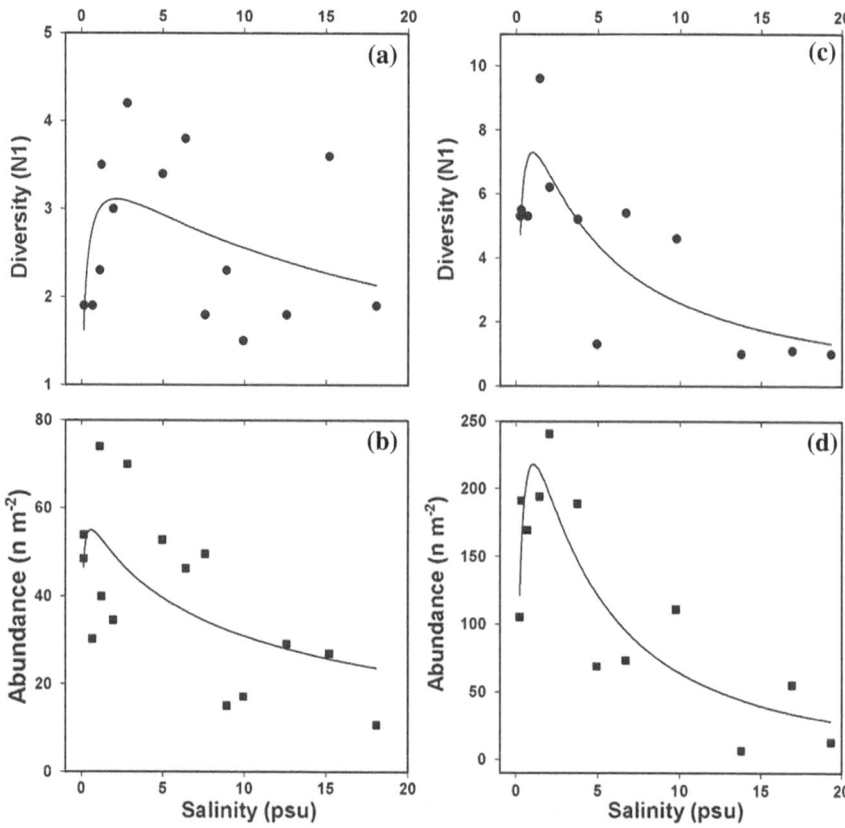

**Fig. 3.14** Relationship between total mollusc diversity (**a**) and abundance (**b**) versus salinity at Myakka River and diversity (**c**) and abundance (**d**) versus salinity at Peace River. *Circles* Hill's N1 diversity index; *squares* abundance. From Montagna et al. (2008a)

The mollusk species *Corbicula fluminea, Rangia cuneata, Polymesoda caroliniana, Tagelus plebius* all exhibited significant bell-shaped responses to salinity, with *C. fluminea* abundance peaking at 0.6 psu, *R. cuneata* at 3.7 psu, *P. caroliniana* at 4.9 psu and *T. plebius* at 7.3 psu (Fig. 3.15). These salinity ranges can be used to predict changes in mollusk assemblages in response to alterations in salinity that result from actual or simulated changes in freshwater inflow.

Mollusk species appeared to be controlled more by water column hydrography than by sediment composition, with salinity as the most important environmental variable and proxy for freshwater inflow.

### 3.2.5  St. John's River Macroinvertebrates

An analysis was conducted using existing data in the St. John's River to investigate the relationship between freshwater inflows and estuarine macrobenthic communities to support water supply planning (SJRWMD 2012). The analysis

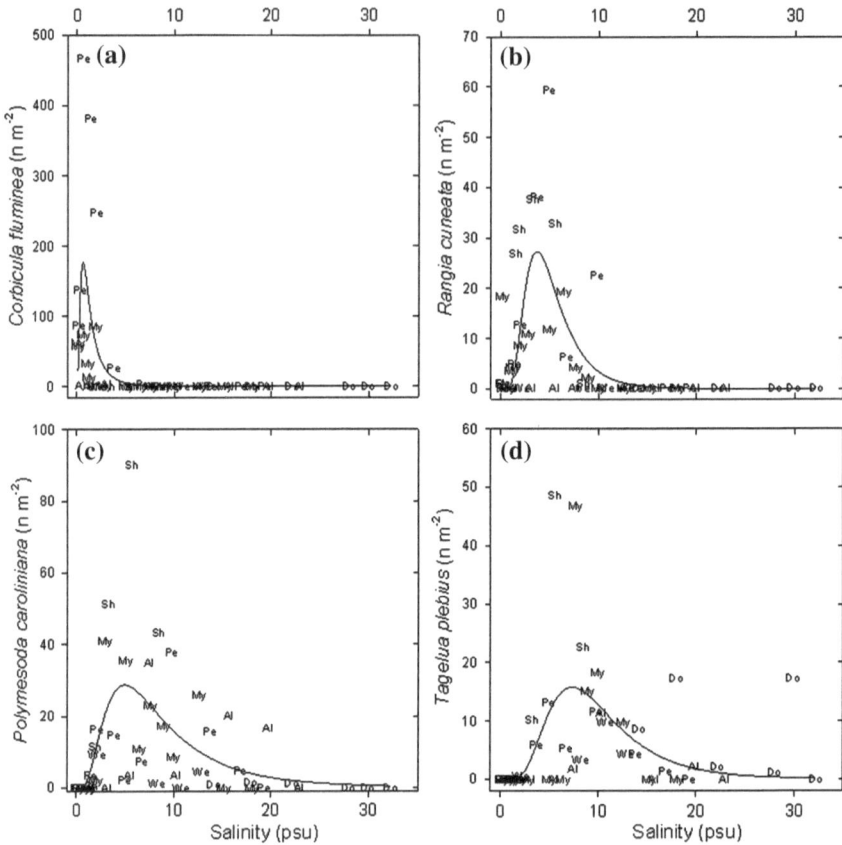

**Fig. 3.15** Mollusk species abundances versus salinity in southwestern Florida estuaries (from Montagna et al. (2008a). **a** Corbicula fluminea; **b** Polymesoda caroliniana; **c** Rangia cuneata; and **d** Tagelus plebeius. Symbol key: Al, Alafia River; Cu, Curry Creek; Do, Dona/Roberts Bay; My, Myakka River; Pe, Peace River; Sk, Shakett Creek; Sh, Shell Creek; We, Weeki Wachee River

approach was similar to the meta-analysis of southwest Florida mollusk data (Montagna et al. 2008a). The spatial scale of the study included eight SAV stations within the St. John's Estuary (Fig. 3.16).

Macrofaunal abundance and diversity peaked when salinity was 1.5 and 1.0 psu, respectively (Fig. 3.17). The rate of change of abundance and diversity with salinity was greater when salinities were below 1.5 and 1.0 psu, respectively than when salinities were above these values. The correlations between salinity and both abundance and diversity were high but not significant.

Although the salinity ranges are small in the study area, specific species and taxa were found that are indicators of the various salinity regimes (Table 3.4). Diversity increases as salinity increases. No specific bivalve or polychaete indicator was found in the fresh water zone (such as Crescent Lake). Only one indicator, the bivalve

**Fig. 3.16** Location of sample sites along the St. John's River, Florida, USA (from Montagna et al. 2008b)

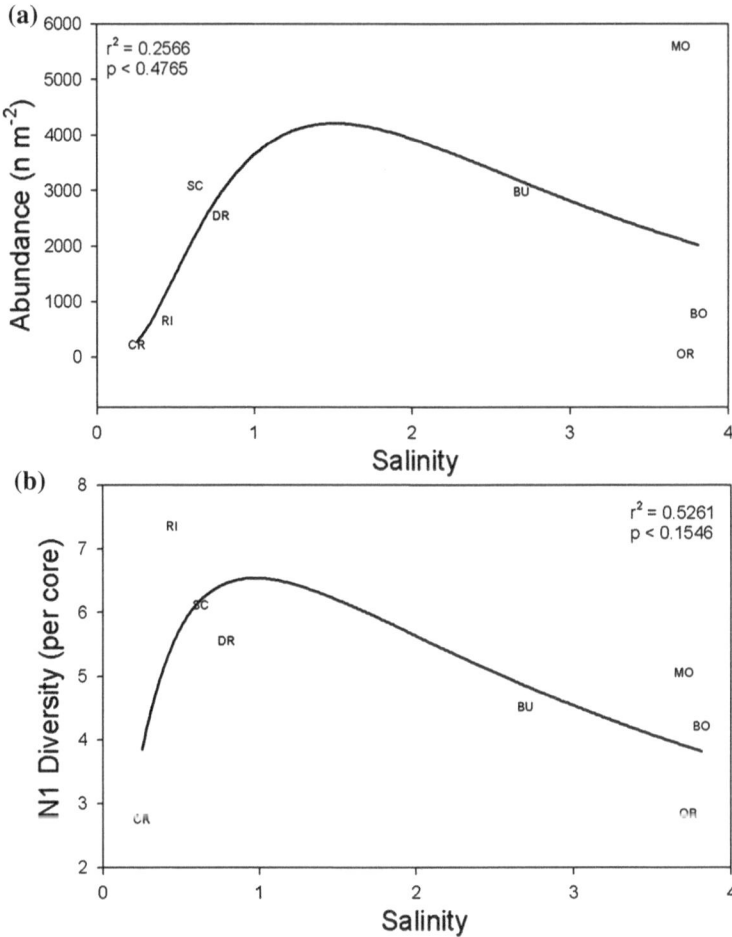

**Fig. 3.17** Salinity versus **a** Abundance and **b** Hills N1 diversity in the St. John's Estuary

*Rangia cuneata*, was an indicator for the fresh oligohaline zone (such as Rice Creek and Scratch Ankle Creek). But there were three additional species that were found to be indicators of the salty oligohaline zone (found in the remaining stations).

There was a clear correlation between salinity and macrofaunal characteristics at these eight stations. The macrofaunal community structures of each station were grouped more by salinity zone then whether the communities are inside or outside of grass beds (Fig. 3.18). Crescent Lake, with lowest mean salinities (0.2 psu has a community structure significantly different than all other communities. The macrofaunal community structures of Rice Creek and Scratch Ankle are similar to each other but significantly different from all others. Rice Creek and Scratch Ankle have the second lowest mean salinities (0.5–0.6 psu). The four stations with the highest salinities (2.4–4.2 psu) have similar macrofaunal community structure

**Table 3.4** St. John's River oligohaline macrobenthic indicators (*B* bivalves, *P* polychaetes)

| | Salinity zone | | |
| --- | --- | --- | --- |
| | 0.3 psu | 0.4–0.6 psu | 2.0–3.9 psu |
| Indicators | Crescent lake | Rice creek/Scratch ankle | All other stations |
| Species | | *Rangia cuneata* (B) | *Marenzelleria viridis* (P) |
| | | | *Laeonereis culveri* (P) |
| | | | *Polymesoda caroliniana* (B) |
| | | | *Rangia cuneata* (B) |
| Higher Taxa | No bivalves | | More polychaetes |
| | | | More bivalves |

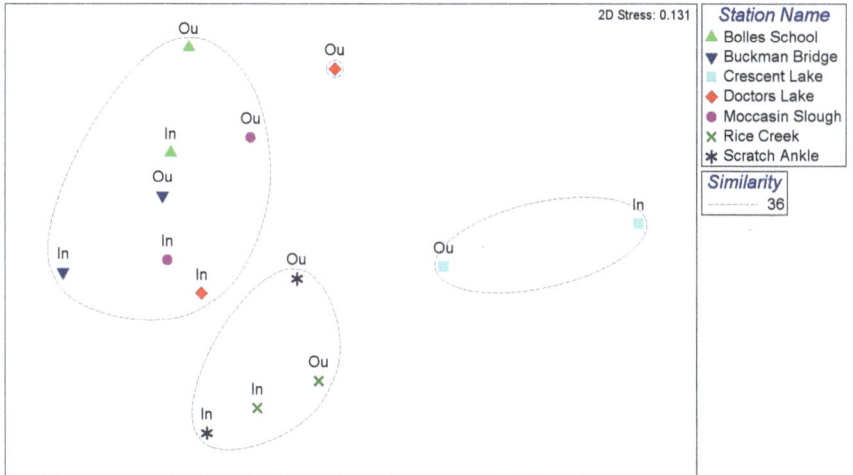

**Fig. 3.18** Nonmetric Multidimensional Scaling plot of SAV stations labeled by station and substrate type; inside (*In*) and outside (*Ou*) grass beds. Similarity contours from cluster analysis are overlaid

except for the Doctors Lake macrofaunal community that is found outside the grass beds. There was a greater change in community structure with salinity when salinities were below 1 psu then when salinities are between 2 and 4 psu.

There was also a greater change in macrofaunal abundance and diversity at lower salinities than higher ones. Macrofaunal abundance and diversity increase sharply to peaks at 1.5 and 1.0 psu, respectively, before slowly decreasing as salinity increases. The variability of macrofaunal abundance and macrofauna around the correlation curves created indicates that other variables in addition to salinity influence macrofaunal communities.

The investigation established that there is a clear change in macrofauna communities with salinity. However, results are based on the limited salinity range within which the submerged aquatic vegetation (SAV) stations are located. If additional macrofauna and salinity data from a wider salinity range further downstream of the

existing SAV stations were incorporated with the existing data, then the results of the
nonlinear model that determined the optimum salinities would likely be very different.


## 3.3 Tri-State Water Sharing Issues
## Within the Apalachicola-Chattahoochee-Flint Basin

The Apalachicola River is the largest river by volume in Florida, and the 4th largest river in the southeastern United States (FL DEP 2010). The river is formed
by the confluence of the Chattahoochee and Flint Rivers at the Florida–Georgia
Border (Fig. 3.19). The Apalachicola River discharges into Apalachicola Bay,
one of the most productive estuarine systems along the US Gulf of Mexico coast.
The Apalachicola-Chattahoochee-Flint (ACF) basin is 51,800 km$^2$ in area and is
shared by Georgia, Florida, and Alabama (Lin 2010). Land use is primarily agriculture within the state of Georgia and silviculture within Florida.

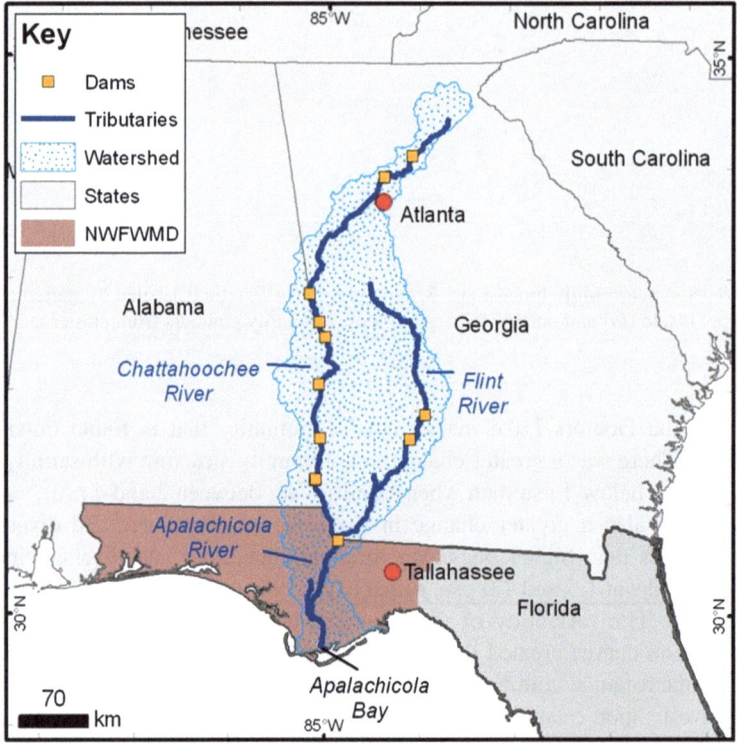

**Fig. 3.19** Location of Apalachicola-Chattahoochee-Flint River system and drainage basin in
Georgia, Alabama, and Florida, USA

**Table 3.5**   Timeline of Apalachicola-Chattahoochee-Flint water use issues

| Years | Action |
|---|---|
| 1946 | Congress authorizes creation of five dams (four on Chattahoochee River, one on Apalachicola River) in an effort to develop the nation's waterways |
| 1956 | Buford Dam completed, creates Lake Sydney Lanier |
| 1957 | Jim Woodruff Dam completed, creates Lake Seminole |
| 1963 | Walter F. George Dam completed, creates Lake Eufala; Andrews Dam (navigation only) completed, creates Lake Andrews |
| 1975 | West Point Dam completed, creates West Point Lake |
| 1988 | Apalachicola Bay declared a federal disaster area after reduced flow in the Apalachicola-Chattahoochee-Flint Rivers damages oyster harvest |
| 1990 | Florida and Alabama sue Georgia and US Army Corps of Engineers |
| 1992 | Governors of Georgia, Florida, and Alabama (and the US Army Corps of Engineers) sign a 5-year memorandum of agreement (MOA), puts the lawsuit on hold; States initiate Comprehensive Study of Resources in the ACF basin as directed by MOA |
| 1997 | Apalachicola-Chattahoochee-Flint interstate compact negotiated, approved by congress |
| 1998 | Deadline for compact: December 1998, but extended 12 times |
| 2003 | Compact expired with no settlement. Lawsuit reactivated, focus on endangered species act |
| 2007 | Governors of Georgia, Florida, and Alabama meet with President Bush regarding drought in southeastern US, develop exceptional drought operations (EDO) plan |
| 2008 | US Army Corps of Engineers begins implementing EDO plan |
| 2010 | Litigation continues; Georgia, Florida, and Alabama file motions to make all settlement negotiations confidential |

There has been a long history of negotiation and litigation between Florida, Georgia, Alabama, and the US Army Corps of Engineers over water sharing issues within the ACF basin (Table 3.5). As of July 2009, there were seven separate lawsuits (FL DEP 2010). At the center of the debate is the US Army Corps' operation of four major hydropower dams on the Chattahoochee River: Buford Dam (north of the city of Atlanta), Jim Woodruff Dam (at intersection of Chattahoochee and Flint Rivers), West Point Dam (south of the city of Atlanta), and George Dam (north of Woodruff Dam). Lake Sydney Lanier, on the Chattahoochee River near Atlanta, is the largest reservoir on the system, providing 65 % of total conservation storage (Wilber 1992). The conflict stems from three main issues: (1) urban and agricultural development within Georgia that has increased upstream water withdrawals, (2) reduced instream flows to Florida affecting the productivity of fisheries in Apalachicola Bay, (3) Alabama's concerns about downstream and cumulative impacts of water resource actions (Lin 2010; NWFWMD 2010).

The water use needs of Georgia, Alabama, and Florida have primarily been met by the construction of dams and reservoirs. Georgia dominates consumptive water use in the ACF basin with municipal and industrial uses exceed 1 million $m^3$ per day (Carter et al. 2008). In addition, Georgia's agricultural uses of ground and surface waters that affect ACF flows can exceed 2.4 million $m^3$ per day during dry summer growing seasons. Direct withdrawal of water from federal ACF reservoirs for agricultural water supply is not authorized. The city of Atlanta,

Georgia obtains 99 % of its water from surface water supplies, with 72 % of the water derived from Lake Lanier and the Chattahoochee River. Alabama's water use needs from the ACF basin are considerably less than Georgia's. Alabama averages less than 190,000 m$^3$ per day, primarily for municipal and industrial uses (Carter et al. 2008). Florida withdraws less than 38,000 m$^3$ per day from the Apalachicola River. In 2006, the Northwest Florida Fisheries Management District reserved most of the flows from the Apalachicola River and one of its tributaries, the Chipola River, for protection of fish and wildlife in Apalachicola Bay.

Although dam construction in the ACF basin was primarily for the purposes of power generation, navigation and flood control, the dams have also become important for recreation, water quality, and protection of fish and wildlife resources (Carter et al. 2008). In 1998, a Draft Environmental Impact Statement (DEIS) was prepared focusing on the potential impacts of surface water allocation within the ACF basin in Alabama, Georgia and Florida, including Apalachicola Bay (DEIS 1998). In particular, the DEIS examined potential effects of water allocation on water, biological, socioeconomic, and cultural resources from 1995 to 2050. The DEIS provided a summary of potential impacts of low, moderate, and high river flow scenarios on each major resource area (Table 3.6).

Freshwater inflows to Apalachicola Bay help support a productive coastal ecosystem including a variety of commercial and recreational fisheries. In particular, the oyster fishery in Apalachicola Bay provides over 90 % of Florida's oyster harvest and 10 % of the US oyster harvest (Andree 1983). Freshwater inflows are important to the maintenance of the oyster fishery. A 1992 study demonstrated that low river flows in the ACF system were positively correlated with oyster catch per unit effort (CPUE) 2 years later, indicating that increased bay salinities allow

**Table 3.6** Potential impacts of surface water allocation on water, biological, socioeconomic and cultural resources

| Resource | Potential impacts |
| --- | --- |
| Water | During moderate and high flow years, water quantities should be sufficient to meet demands. During low-flow years, river flows would be substantially reduced and reservoir levels maintained near full pool. Navigation flows would be reduced slightly |
| Biological | Impacts on terrestrial vegetation, floodplain freshwater wetlands would be minimal. Estuarine wetlands would experience greatest impacts in low-flow scenarios. Wildlife expected to have only limited habitat loss for brief periods during low-flow periods. Reduced inflow to Apalachicola Bay increases salinity, predatory oyster drill abundance, and oyster mortality |
| Socioeconomic | Population growth is expected to follow water availability. Low and moderate flow periods will create less desirable conditions for recreational activities such as wade fishing, canoeing and rafting. Municipalities at the headwaters of the ACF basin will experience substantially greater adverse impacts than areas downstream. No water supply shortages for irrigation are predicted |
| Cultural | Three potential effects of water level management may affect cultural and archeological resources: erosion, deposition of sediment, and access. Of the three flow scenarios, the low-flow scenario would produce downstream flows with lower velocity, resulting in fewer opportunities for erosion |

increased predation on newly settled spat, thus reducing the population of marketable oysters 2 years later (Wilber 1992). From a socioeconomic perspective, more than 1,000 people are employed by the oyster industry in Florida's Franklin County (Carter et al. 2008).

Dams and reservoirs alter the natural flow regimes of rivers and streams, fragment the free-flowing riverine system, alter physical and chemical processes, and disrupt biological communities (Cowie 2002). Recent attention has been focused on the role of freshwater from the ACF basin in protecting and maintaining threatened and endangered species. Because of the habitat alteration caused by dams and reservoirs, 34 fish species and 16 freshwater mussel species are imperiled in Georgia (Cowie 2002). Particular attention has been focused on four species listed under the US Endangered Species Act: the Gulf sturgeon *Acipenser oxyrinchus desotoi*, fat threeridge mussel *Amblema neislerii*, Chipola slabshell mussel *Elliptio chipolaensis*, and purple bankclimber mussel *Elliptoideus sloatianus*. All four species are susceptible to changes in water flow, temperature, and dissolved oxygen levels (Carter et al. 2008). In addition, the Gulf sturgeon is anadromous and needs to migrate upstream from the Gulf of Mexico to spawn. The three mussel species live in sand and gravel stream bottoms. Dams alter their free-flowing water habitats and restrict larval dispersal, resulting in small and isolated populations (Carter et al. 2008).

Drought conditions plagued much of the southeastern US beginning in 2006. By mid-2007, only one-third of the system conservation storage was left in ACF reservoirs (Zeng et al. 2009). In response, the states of Georgia, Florida, and Alabama, and the US Army Corps of Engineers developed an Exceptional Drought Operations plan (EDO). The EDO plan replaced all previous flow provisions with a minimum flow requirement of 134.5 $m^3$ $s^{-1}$. All flows above this level could be stored in the ACF reservoirs. The basic idea was to maintain water in ACF reservoir storage to protect against further drought. Maintaining low flows was predicted to cause less harm to species during drought years than if the water was completely depleted.

The issue of shared water in the ACF basin is far from resolved, and the coming year will be filled with intense activity as litigation continues among Georgia, Florida, and Alabama at the federal level. Emerging issues of concern will include future climate change and clean water act jurisdiction and the impacts of these issues on water sharing and biological resources.

## 3.4 California, USA

California has the third largest area of all the states in the United States, covering approximately 414,000 $km^2$. California is also the most populous in the USA, with an estimated 37 million people (US Census Bureau 2010). The state is highly variable in terms of climate and precipitation. Seasons in California are either wet (October to March) or dry (April to September). Two-thirds of the average annual precipitation falls in the northern third of the state, and the eastern side of California's mountain ranges create rain shadow deserts with extremely dry climates (Fig. 3.20).

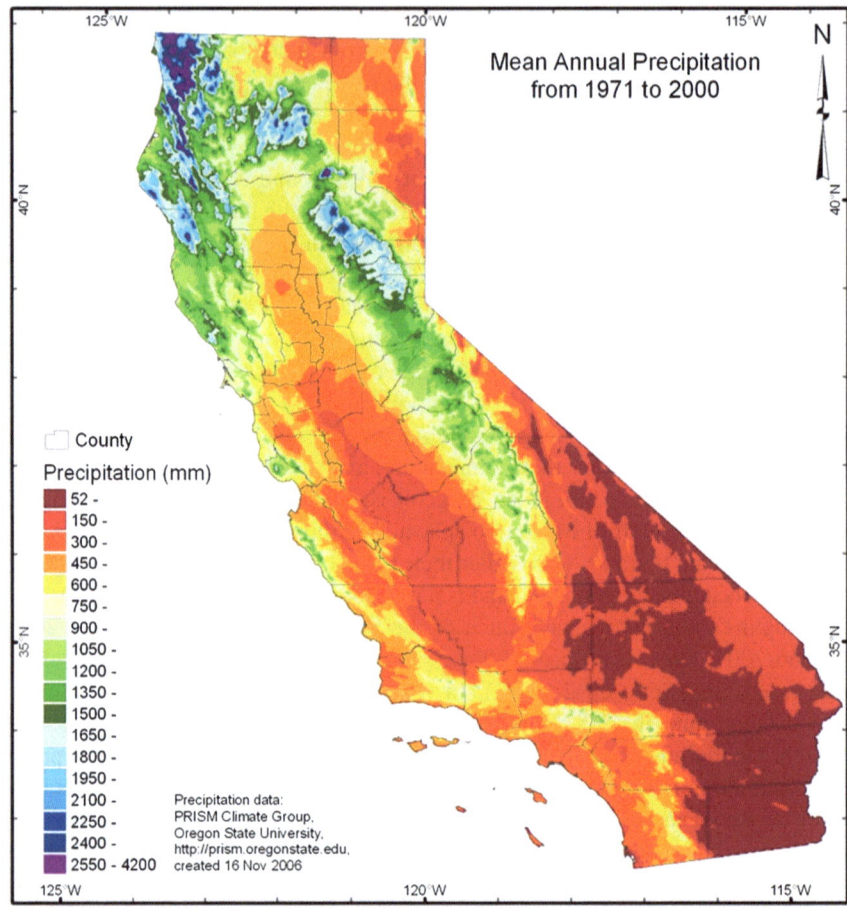

**Fig. 3.20**   Rainfall gradient in California. Average annual precipitation from 1971 to 2000

The regulation, protection, and administration of water quality in California are carried out by the State Water Resources Control Board and nine Regional Water Quality Control Boards (Fig. 3.21). These groups are tasked with protecting and enhancing the quality of California's waters for present and future generations (State Water Resources Control Board 2010). The State Water Resources Control Board is one of five departments within the California Environmental Protection Agency and is responsible for allocating surface water rights in California, as well as setting statewide policy and regulations for water quality control (State Water Resources Control Board 2010). The State Board comprises five full-time members appointed by the governor for 4-year terms.

Because of regional differences in water quality and quantity, the state is divided into nine regions for administration of California's water quality control program: (1) North Coast, (2) San Francisco Bay, (3) Central Coast, (4) Los

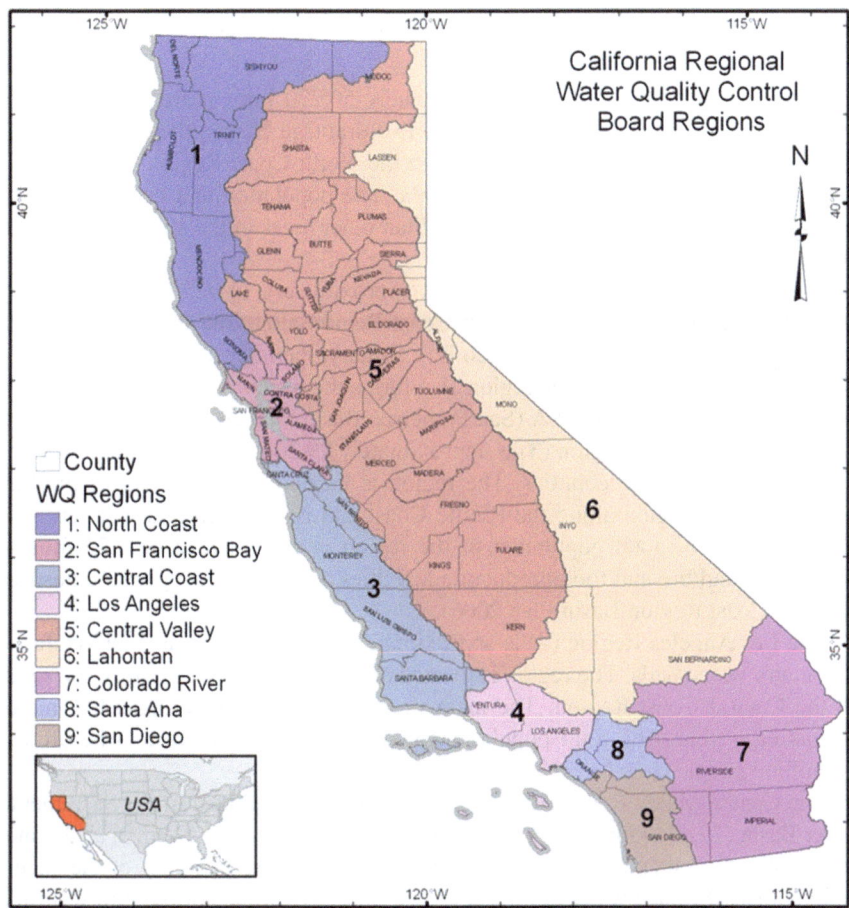

**Fig. 3.21** Location of the nine California Regional Water Quality Control Boards

Angeles, (5) Central Valley, (6) Lahontan, (7) Colorado River Basin, (8) Santa Ana, and (9) San Diego (Fig. 3.21) (State Water Resources Control Board 2010). Each Regional Board is responsible for local implementation of policy and regulations, and development of water quality decisions for its region. The Regional Boards each consist of nine part-time members appointed by the governor for 4-year terms. Each Regional Board has developed a Basin Plan (or Water Quality Control Plan) for their region. The Basin Plans are developed to preserve and enhance water quality and protect the beneficial uses of regional waters (State Water Resources Control Board 2010). These plans are designed to be more than an abstract set of goals and policies; rather they set forth water quality standards for surface and ground waters, identify water quality problems and recommend control measures (Lahontan Basin Plan 1995). The Basin Plans also serve to implement the US Clean Water Act in California.

## 3.4.1 Regional Water Quality Control Board Profiles

The North Coast Region (1) is approximately 50,220 km$^2$ in area in northwestern California. The region is divided into two drainage basins, the Klamath River and North Coastal Basins. Precipitation in the North Coast Region (64.1 km$^3$) is greater than any other region in California. The population and economy of the basin have developed much more slowly than other areas in the state (North Coast Basin Plan 2007).

The San Francisco Bay Region (2) is approximately 11,922 km$^2$ in area and is located directly south of the North Coast Region along the Pacific Coast. The dominant feature is the San Francisco Bay Estuary, the largest estuary on the west coast of the United States. The region also forms the backdrop of the fourth largest metropolitan region in the USA (San Francisco Basin Plan 2007).

The Central Coast Region (3) is approximately 29,200 km$^2$ in area and includes all or portions of nine counties. The region is located directly south of the San Francisco Bay Region along the Pacific Coast and is considered arid in terms of precipitation. The total population of the region is estimated to be 1.22 million people. Agriculture and food production activities are the major regional industries (Central Coast Region Basin Plan 2006).

The Los Angeles Region (4) is approximately 10,130 km$^2$ in area and includes five major watersheds. The region is located along the Pacific Coast, directly south of the Central Coast Region. The main precipitation occurs during a few major storms each year. The total population of the region is estimated to be around 12 million people (Los Angeles Region Basin Plan 1995).

The Central Valley Region (5) covers the entire area included in the Sacramento River Basin (70,474 km$^2$) and the San Joaquin River Basin (41,129 km$^2$), about one-fourth of California's total area. These rivers provide more than half of the state's water supply. Surface waters from these two drainage basins converge to form the Sacramento-San Joaquin River Delta, which drains into San Francisco Bay. The Central Valley Region forms the eastern (inland) boundary of the North Coast, San Francisco Bay, and Central Coast Regions (Central Valley Region Basin Plan 1998).

The Lahontan Region (6) is approximately 85,809 km$^2$ in area and includes both the highest (Mount Whitney) and lowest (Death Valley) points in the continental United States. The region is located in a rain shadow, but precipitation in the higher elevations can be high (up to 178 cm yr$^{-1}$). The Lahontan Region covers almost the entire eastern edge of the state and is divided into 12 major watersheds. The region has relatively low resident population and an agricultural emphasis on livestock grazing rather than crops (Lahontan Basin Plan 1995).

The Colorado River Region (7) covers approximately 51,800 km$^2$ in southeastern California and includes all or part of four counties. The Colorado River is the most important waterway in the region, which contains a small portion of the total Colorado River drainage area. The most significant feature in the region is the Salton Trough, an extension of the Gulf of California that contains the Salton Sea. The region has relatively low population density (Colorado River Basin Plan 2006).

The Santa Ana Region (8) is the smallest of the nine water quality control board regions in the state (approximately 7,250 km$^2$). The region located along the Pacific coast directly south of the Los Angeles Region, with its eastern boundary formed by the Colorado River Region. The most significant hydrologic feature in the region is the Santa Ana River. Although geographically small, the estimated population of the region is approximately 4 million people (Santa Ana Basin Plan 2008).

The San Diego Region (9) covers approximately 10,100 km$^2$ in the southwest corner of California. The region is located directly south of the Santa Ana Region along the Pacific Coast and is bounded on the eastern edge by the Colorado River Region. The region is divided into 11 major watersheds. The estimated population of the San Diego Region is approximately 9 million people (San Diego Basin Plan 1994).

## 3.4.2 San Francisco Bay

The San Francisco Bay Estuary on the central coast of California is the largest estuary on the west coast of the United States, roughly 2,575 km$^2$ in area (Fig. 3.22). Almost all the freshwater to the estuary is supplied by the Sacramento and San Joaquin Rivers, which drain via the Sacramento-San Joaquin Delta into the estuary and mix with the saline waters of the Pacific Ocean. The estuary is commonly divided into four subregions: Suisun Bay, North Bay/San Pablo Bay, Central Bay, and South Bay. Freshwater inflow to the San Francisco Bay Estuary supports a wide range of habitats including deepwater channels, tidelands, marshlands, and freshwater streams (San Francisco Bay Basin Plan 2007). These habitats in turn support diverse communities of crabs, clams, fish, and birds, as well as migrating waterfowl. However, much of the freshwater is captured upstream by the dams, canals, and reservoirs that are part of California's water diversion projects.

California has been dealing with water allocation issues for decades. The Sacramento-San Joaquin River Delta is the largest source of California's freshwater, providing water to approximately 23 million people, or two-thirds of all Californians (CALFED 2010). The Delta drains approximately 100,000 km$^2$, or 40 % of the area of California (Kimmerer 2002). Precipitation falls mainly to the north of the Sacramento-San Joaquin Delta in winter and spring, but the greatest demand for freshwater is in the south during the summer. During the dry summer and fall months, the Sacramento-San Joaquin River Delta system is used to move water from water-rich reservoirs in the north to water-poor farms and cities to the south (Kimmerer 2002). Water is exported for irrigation and municipal use by large diversion pumps operated by the Central Valley Project and the California State Water Project. These facilities represent one of the world's largest man-made water systems (Nichols et al. 1986). Since 1977, about half of the total possible flow into the estuary has been diverted, with about 80 % used for agricultural irrigation, and the rest used for municipal and industrial purposes (Pierson et al. 2002).

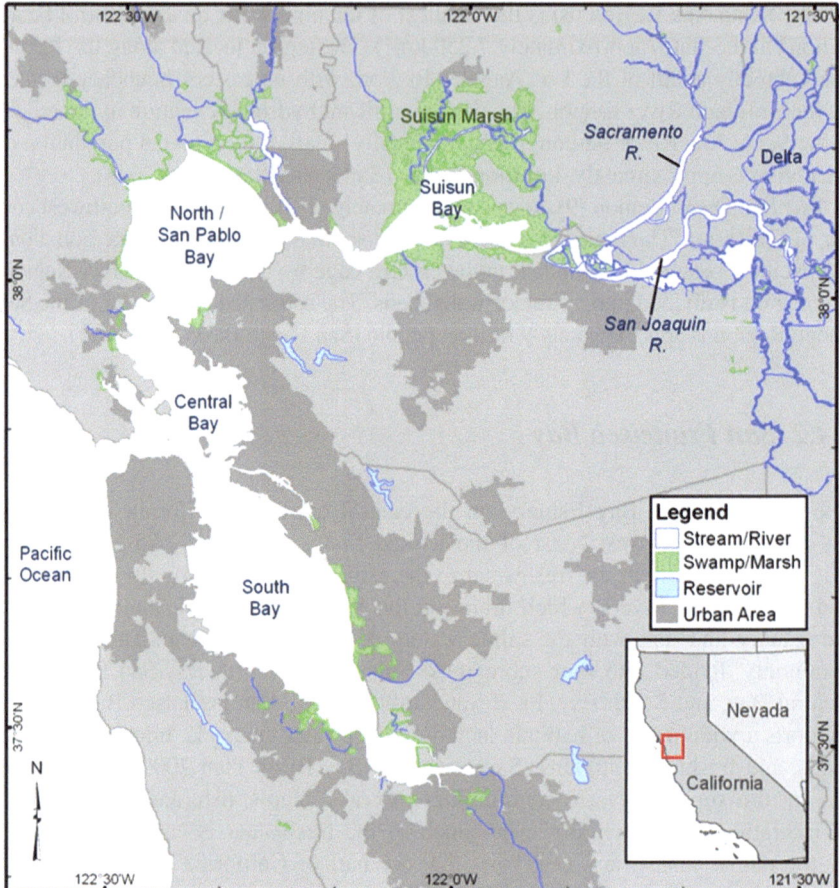

**Fig. 3.22**  San Francisco Bay Estuary, California, showing the different subregions (Suisun Bay, North Bay/San Pablo Bay, Central Bay, and South Bay)

Freshwater diversion has altered the hydrograph more through a change in seasonal patterns than in total flow (Kimmerer 2002). Water is released from reservoirs to prevent salt water intrusion into the Sacramento-San Joaquin Delta from the Pacific Ocean. Both the physical process of diverting water and the resulting alterations in flow have affected biological communities of the San Francisco Bay Estuary (Nichols et al. 1986). In particular, concerns have increased over the effects of altered freshwater flows on several fish species, including the delta smelt (*Hypomesus transpacificus*), threadfin shad (*Dorosoma petenense*), longfin smelt (*Spirinchus thaleichthys*) and striped bass (*Morone saxatilis*) (Feyrer et al. 2007; Sommer et al. 2007). The delta smelt is endemic to the low-salinity habitats of the northeastern San Francisco Bay Estuary. Although the species was harvested commercially in the nineteenth and twentieth century, recent declines have resulted in its listing as threatened under the US Endangered Species Act (Bennett 2005).

Another ecological consequence of altered freshwater flows is the absence of a summer phytoplankton bloom in northern San Francisco Bay during extremely low inflows, which can suppress the pelagic food web, fish recruitment, and fisheries yields (Nichols et al. 1986).

The State Water Resources Control Board adopted an inflow standard based on a desired salinity gradient within the bay. Freshwater inflow to the San Francisco Bay Estuary is regulated in part using the variable X2. This variable represents the longitudinal position in the estuary of the 2 psu salinity isohaline, measured 1 m from the bottom of the water column, averaged over a suitable time period (>1 day) (Pierson et al. 2002). This low-salinity zone moves primarily in response to flow (i.e., located further upstream during low-flow periods). The direction and magnitude of fluctuations is measured as the distance in kilometers from the Golden Gate Bridge. When freshwater inflows to the estuary are high, X2 is located seaward in direction (the value of X2 is lower). The standard was chosen as an indicator of estuarine habitat and significant relationships have been found between X2 and phytoplankton, shrimp, mysids, and desirable fish larvae (Kimmerer 2002). This approach is important because the rule can be linked to many ecosystem components and all trophic levels within the estuary. California's regulations allow the salinity guideline to change seasonally.

Dissatisfaction with water resource management led to the establishment in 1995 of the CALFED Bay/Delta Science Program. The mission of the joint federal-state program is to provide the best possible scientific information for water and environmental decision making in the San Francisco Bay/Sacramento-San Joaquin River Delta system (CALFED 2010). In November 2009, the California legislature passed a comprehensive water package to fund water issues including Delta sustainability, conservation and watershed protection, and water recycling and conservation programs.California Senate Bill SBx7 1 created a new Delta Stewardship Council, an independent state agency tasked with developing a Delta Plan by January 2012 with goals of providing a more reliable water supply and protecting and restoring the Delta ecosystem. Under the same bill, the CALFED Bay/Delta Science Program became the Delta Science Program, reporting to the new Council (Delta Council 2010).

As part of the recent legislation, SBx7 1 Division 35: the Sacramento-San Joaquin Delta Reform Act of 2009, the California State Water Resources Control Board was tasked with developing new flow criteria for the Delta ecosystem necessary to protect public trust resources. The flow criteria must include the volume, quantity, and timing of water necessary for the Delta ecosystem under different conditions. The compliance deadline for development of these criteria is 9 months. In addition, the California Department of Fish and Game, with assistance from the US Fish and Wildlife Service and the National Marine Fisheries Service, was tasked with developing and recommending flow criteria and quantifiable biological objectives for species of concern that are dependent on the Delta. The deadline for these recommendations is 12 months. Thus, flow criteria and recommendations should be forthcoming sometime in mid-late 2010. The comprehensive management plan for the delta (Delta Plan) will be completed by January 1, 2012.

## 3.5  Gulf of California, Mexico

Historically, the Gulf of California received large volumes of freshwater from the Colorado River. The Colorado River forms the southeastern border of California and Arizona (Fig. 3.23). The watershed includes parts of seven US states and two Mexican states. The Colorado River is the primary water source for the southwestern United States, irrigating more than 14,900 km$^2$ of farmland and supplying water to nearly 30 million people (Luecke et al. 1999). The natural course of the river flows from the headwaters in northwestern Colorado south and west through Utah, Nevada, Arizona, California, and into Mexico en route to the Gulf of California. River flows are highly variable, ranging from maximum annual natural flows of 30.5 km$^3$ yr$^{-1}$ to minimum annual natural flows of 6.2 km$^3$ yr$^{-1}$ (California's Colorado River Water Use Plan 2000).

**Fig. 3.23**  Colorado River, stretching through the southwestern US states of Colorado, Utah, Arizona, Nevada and California en route to Mexico and the Gulf of California (from Varady et al. 2001)

The Colorado River has a long history of water rights negotiations. The statutes, court decisions and decrees, contracts, interstate compacts, and administrative rules and regulations that define the apportionments of the Colorado River to US states and Mexico are collectively known as "the Law of the River" (Varady et al. 2001). Most significantly, the Colorado River Compact of 1922 was negotiated among seven states in the basin of the Colorado River to govern water allocation. For the purposes of the compact, the river basin is divided into an Upper Basin that includes Colorado, Utah, Wyoming, New Mexico and Arizona, and a Lower Basin that includes California, Arizona and Nevada. The compact apportioned 9.25 km$^3$ of water per year to each basin (Colorado River Compact 1922). In 1944, the United States also signed a treaty agreeing to deliver an annual quantity of 1.85 km$^3$ of water per year to Mexico (Treaty Series 994 1944). The amount of water allocated was based on the expectation that the river's average flow was 20.2 km$^3$ of water per year. Unfortunately, the long-term average river flow is considerably less (approximately 18.0 km$^3$ yr$^{-1}$). As a result the sum of the compact apportionments and the Mexican treaty exceed the flow of the Colorado River in most years (Woodhouse et al. 2005) (Table 3.7). Because of intense human alteration in the form of dams and reservoirs, little or no water has reached the Gulf of California since 1960.

Historically, the Colorado River Delta comprised a vast area (7,810 km$^2$) of riparian, freshwater, brackish and tidal wetlands, and served as an important migratory bird stopover and nursery for endemic marine species (Luecke et al. 1999). However, decades of dam and irrigation projects have diverted water from the Colorado River for human use, resulting in a much reduced delta of small wetlands and brackish mudflats. In the 1930s, following the construction of the Hoover Dam, virtually no freshwater reached the delta for 6 years as the Lake Mead reservoir filled behind the dam. From 1963 to 1981, a similar situation occurred following the construction of the Glen Canyon Dam and subsequent filling of the Lake Powell reservoir. Floodwaters are released only when the US Bureau of Reclamation predicts flows that exceed the system's capacity for use and storage

**Table 3.7**  Colorado River allocations

| Allocation | Km$^3$ yr$^{-1}$ |
|---|---|
| *Upper basin* | |
| Colorado | 4.8 |
| Utah | 2.1 |
| Wyoming | 1.2 |
| New Mexico | 1.0 |
| *Lower basin* | |
| Arizona | 3.5 |
| California | 5.4 |
| Nevada | 0.4 |
| *Additional allocations* | |
| Mexico | 1.9 |
| Total allocations | 20.3 |
| Long-term average flow | 18.0 |

(Luecke et al. 1999). During the twentieth century, river flows to the Colorado River Delta have been reduced nearly 75 % (Fradkin 1981; Glenn et al. 1999).

Reductions of Colorado River flow to the delta have resulted in significant physical and biological impacts. On the physical side, reduced delivery of silt and nutrients as well as higher salinity and pollutant concentrations have been recorded (Luecke et al. 1999). In addition, the delta has taken on the highly unusual condition where erosion (versus accretion) is the dominant physical process (Thompson 1968). Biologically, organism abundance is linked to Colorado River flow. The relative abundance of postlarval blue shrimp *Litopenaeus stylirostris*, which constitute the dominant shrimp fishery, was significantly higher when flow from the Colorado River reached the upper Gulf of California (Aragón-Noriega and Calderón-Aguilera 2000). Declines in the catch of *Totoaba macdonaldi*, a once-abundant and now-endangered fish species in the Gulf of California coincide with increased fishing effort and decreases in Colorado River flow (Cisneros-Mata et al. 1995). During historical periods of natural river flow, conservative estimates of bivalve mollusk abundances are around $6 \times 10^9$ (density ~50 m$^{-2}$), whereas the present abundance is ~94 % lower, or around 3 m$^{-2}$ (Kowalewski et al. 2000). The delta is home to the largest populations of two endangered species, the desert pupfish (*Cyprinodon macularius*) and the Yuma clapper rail (*Rallus longirostris yumanensis*) (Glenn et al. 1996). Future reductions in flow could have strong impacts on these and other vulnerable species.

The flashiness of flow conditions can also influence species persistence. In particular, non-native species may not be behaviorally adapted to avoid displacement by sudden floods (Poff et al. 1997). The historical cycles of scour and fill in the Colorado River prevented encroachment of vegetation onto the river banks, whereas recent flow reduction has results in increased wetland and woody vegetation, dominated by the exotic salt cedar tree (*Tamarix* sp.) (Stevens et al. 1995).

At the heart of present-day Colorado River water allocation issues is the fact that "the Law of the River" is based on nineteenth century needs, which differ fundamentally from current needs. However, negotiating water reallocations can be tricky because the gains of one user must be balanced by the losses of another. One option that has been proposed as an environmentally sustainable mechanism for reallocations is to develop a water market system, which would offer several advantages such as: (1) allowing water to be traded on a voluntary basis, (2) allowing water to be directed to the most productive activities thus optimizing the economic returns, and (3) allowing for the creation of water taxes or other allocation systems to ensure that environmental and consumptive uses can be balanced (Varady et al. 2001). However, water markets that need to cross state boundaries involve additional layers of complexity. The water market option will require more research and time to determine if they offer a reasonable solution to interstate and international water allocation issues.

In the past 30 years, partial restoration of river flow has been accomplished by floodwater, periodic reservoir releases, agricultural return flows, and municipal wastewater (Kowalewski et al. 2000). Agricultural drain water has created Cienega de Santa Clara, a 42 km$^2$ *Typha domengensis* marsh containing the largest remaining population of the endangered Yuma clapper rail and many other species of migratory and resident waterfowl (Glenn et al. 2001). Attempts at river restoration have also been conducted via planned flood events. In 1996, a 7-day test flood (approximately 35 % of the pre-dam average spring flood) was released from Glen Canyon Dam. As a result, over 53 % of the beaches increased in size and just 10 % decreased in size (Poff et al. 1997). The flood was of not large enough to significantly reduce populations of nonnative fishes, but studies indicate that similar managed floods can impair nonnative predator and competitor populations, enhancing the survival of native fishes (Valdez et al. 2001). *Totoaba macdonaldi* has experienced an increase in annual survival of in recent years, suggesting recovery of the stock (Cisneros-Mata et al. 1995). It is important to consider the timing and magnitude of planned floods in reference to organism life cycles, biological interactions, and habitat preferences. Although the test flood successfully restored sandbars and was timed to prevent impacts to species of concern, an estimated 10.7 % of the habitat of the endangered Kanab amber snail (*Oxyloma haydeni kanabensis*) and 7.7 % of the snail population were lost to the flood (Stevens et al. 2001).

Involvement of and cooperation between many stakeholder groups is important to the success of any restoration program. Scientists studying Colorado River water issues have interacted with numerous stakeholder groups including federal and state agencies, energy production organizations, Native American tribes, and recreational and environmental organizations (Poff et al. 2003). Identification of consistent funding streams is also important. For example, ecosystem research and monitoring can be supported via allocation of a small portion of the revenues generated at federal hydroelectric power dams. Hydropower revenues from the main dams of the Colorado River Storage Project (Richter et al. 2003) support both the Grand Canyon Monitoring and Research Center and the monitoring element of the Recovery Implementation Program for Endangered Fish Species in the Upper Colorado River Basin (Poff et al. 2003). Monies from environmental damage taxes and mitigation banking may also be developed to secure water or money for the purpose of ecosystem restoration (Pitt et al. 2000).

Large ecosystem restoration is a relatively new practice where science plays an integrative rather than analytic role and the process is generally political (Luecke 2000). In addition, restoration may not equate to reestablishment of the predevelopment ecosystem, but rather the restoration of ecosystem function (Pitt 2001). Thus, establishing metrics of restoration success can be a complex process. For the future, some combination of applied science and management, better-adapted institutions, binational interests, and community-level participation are needed to help resolve the problems facing the Colorado River watershed and Delta (Varady et al. 2001). The United States and Mexico appear to be committed to protect the Colorado River and Delta ecosystems. In 2000, the countries signed an agreement known as Minute 306, which amends the 1944 treaty between the US and Mexico. The agreement seeks to develop joint studies that include possible approaches to ensure use of water for ecological purposes with a focus on defining habitat needs of fish and marine and wildlife species of concern to each country (IBWC 2000). Because the agreement does not specifically dedicate instream flows to the

Colorado River Delta, the future of the delta ecosystem remains in jeopardy. Time will tell whether recent increases in public interest related to Colorado River water issues can drive existing water problems toward resolution in a timely fashion.

## 3.6  Guadiana River Estuary, Europe

The Guadiana River Estuary is located on the Iberian Peninsula, on the southern border between Portugal and Spain. The mesotidal estuary is approximately 22 km$^2$ with an average depth of 6.5 m; tidal amplitudes range from 1.3 to 3.5 m (Faria et al. 2006). The catchment basin for the Guadiana Estuary is the fourth largest in the Iberian Peninsula, at approximately 67,500 km$^2$ (Fig. 3.24). Under the influence of the semiarid Mediterranean climate, the estuary experiences seasonal and interannual variations in natural river inflows and is characterized by severe droughts and heavy floods (Chícharo et al. 2006b). Rainfall is highly variable, with approximately 80 % of precipitation occurring during the fall and winter (Morais et al. 2009).

The Guadiana Estuary has experienced widespread development over the past century. Municipal and agricultural development, particularly since the 1950s–1960s, has steadily decreased the quantity and quality of water reaching the estuary and coastal zone (Chícharo et al. 2001; Rocha et al. 2002; Dias et al. 2004). Freshwater flow reaching the estuary is regulated by more than 100 dams, including the Alqueva dam—Western Europe's largest—in 2002, creating an impoundment with a potential area of 2,500 km$^{-2}$ (Alveirinho et al. 2004). As a result, the volume of water retained in the Guadiana River and not reaching the estuary is approximately 13 km$^3$ yr$^{-1}$ (Dias et al. 2004). Alteration in freshwater discharge has affected the frequency and duration of flooding to the estuary, with a shift to shorter, more abrupt flood events (Chícharo et al. 2006b). The effect is an estuarine system that is freshwater dominated during winter and flood periods and salt wedge dominated during spring and summer (Rocha et al. 2002).

Important ecological benefits have been traditionally provided by the Guadiana Estuary, including nursery function for fish (e.g. *Engraulis encrasicolus*) and shrimp (*Crangon crangon*), and nutrient export to coastal areas, supporting benthic and planktonic food webs and fisheries productivity (Chícharo et al. 2002; Erzini 2005). Preconstruction studies predicted the Alqueva Dam would affect changes in community structure of fish assemblages, alter spawning behaviors, and disrupt migration patterns of early life stages (Chícharo et al. 2003). Postconstruction, community alterations have been documented for a variety of organisms, including plankton and fish (Chícharo et al. 2006a; Faria et al. 2006; Domingues et al. 2007). In addition, first occurrences of several marine invasive species have also been recorded, with potential detrimental effects on native biota (Chícharo et al. 2009).

Wolanski et al. (2006) developed an ecohydrology model for the estuary that suggests Guadiana Estuary ecosystem health requires transient river floods and compromised flow regulation (i.e. via damming). The model predicts strong effects from construction of the Alqueva dam on carnivorous/omnivorous fish, which are now significantly restricted in their distribution in the upper-most region of the estuary. In addition, zooplanktivorous fish and their zooplankton prey are

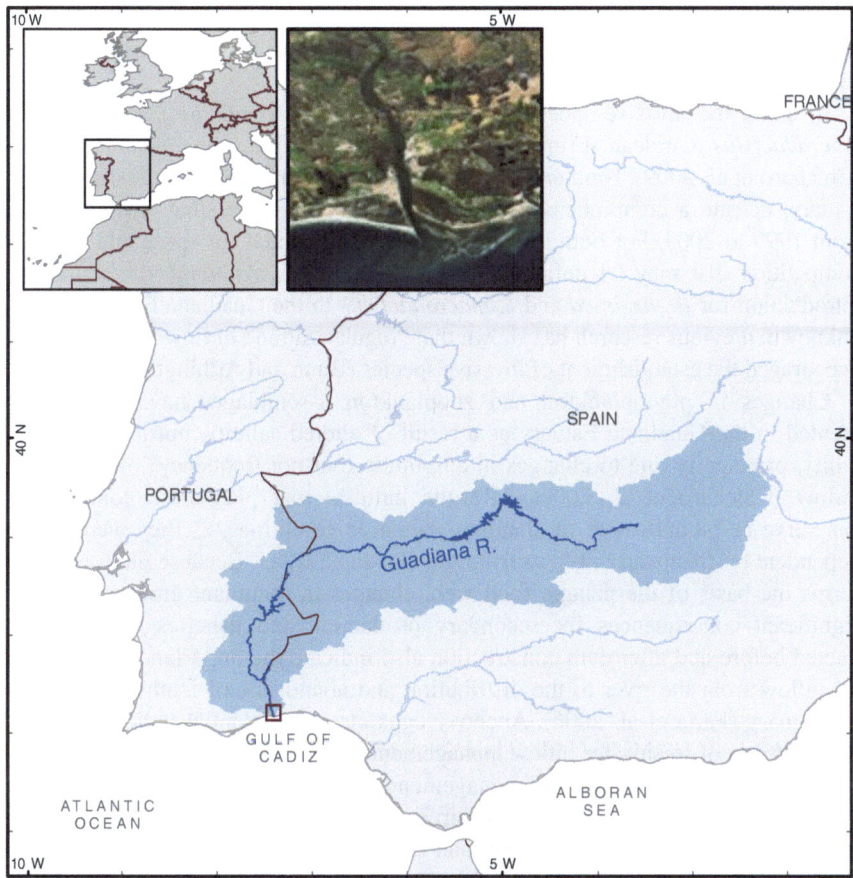

**Fig. 3.24** Guadiana River Estuary, located on the southern border between Portugal and Spain

predicted to decrease in the presence of the dam because their renewal and distribution depend on freshets (Wolanski et al. 2006).

During the construction and filling of the Alqueva dam, freshwater inflow
was the most important factor determining abiotic and biotic variability in the
Guadiana estuary and adjacent coastal area (Morais et al. 2009). The turbidity
maximum zone was displaced toward the upper estuary, 8–16 km further upstream
than previously recorded. Nutrient stoichiometry was also affected. In the upper
and mid-estuary, P limitation was stronger than N limitation year-round, and Si
limitation only occurred near the coast, likely due to changing phytoplankton
dynamics (Morais et al. 2009).

Changes in salinity and seston due to changes in freshwater inflow have also been
shown to have strong effects on the community structure of fish assemblages in the
Guadiana Estuary. Comparing estuarine fish assemblages in the year before (2001)
versus after (2002) construction of the Alqueva dam illustrates marked changes in
species distribution, abundance, and biomass (Chícharo et al. 2006b). Reduction of
freshwater inflow in 2002 due to dam construction and filling was suggested as a

principal factor for reduced spatial separation of freshwater/estuarine/marine fish communities and the upstream displacement of freshwater species during spring and summer 2002 (Chícharo et al. 2006b).

In 2008, the invasive species *Blackfordia virginica* (cnidarian) and *Palaemon macrodactylus* (caridean shrimp) were first observed within the Guadiana Estuary (Chícharo et al. 2009). Neither of these species had been previously detected in the estuary, despite a comprehensive invertebrate species survey that was conducted from 1999 to 2003. For both species, there is the potential for space and resource competition that may be unfavorable to native species. Although the sources of introduction for *B. virginica* and *P. macrodactylus* to the Guadiana Estuary remain unknown, previous research has shown that "regularization" of flow to estuaries has encouraged the establishment of invasive species (Bunn and Arthingthon 2002).

Changes in phytoplankton and zooplankton assemblages have been documented in the Guadiana Estuary as a result of altered salinity, nutrients, and turbidity, principally due to changes in amplitude (but not frequency) of freshwater inflow (Chícharo et al. 2006a). Results indicate that planktonic communities can serve as bioindicators of changing estuarine conditions, as they were highly dependent on freshwater inflows from the Guadiana River. Because phytoplankton forms the basis of the pelagic food web, changes in dominant groups may have significant consequences for secondary production and fisheries. Studies conducted before and after dam construction also indicate the importance of freshwater inflow from the river to the distribution and abundance of ichthyoplankton in the estuary (Faria et al. 2006). Anchovy eggs also are potential indicator species of the effects of freshwater inflow management (Morais et al. 2009). As freshwater pulses can be provided via dam management, use of indicator species can be beneficial for understanding the relationship between the magnitude and frequency of inflow events and estuarine structure and function (Chícharo et al. 2006a). Studies indicate that resource managers should attempt to mimic natural river flow regimes as much as possible, in order to minimize the impact on downstream ecosystems.

## 3.7  Australia

Australia is the driest inhabited continent in the world. Rainfall is also highly variable over space (Fig. 3.25) and time in Australia causing Australia to impound 12 times as much water as would be impounded in similar latitudes in the North America (Cappo et al. 1998a). Throughout Australia, especially in northern regions, a large amount of the rainfall is lost through evapotranspiration. The average population density of the approximately 22 million people in Australia is less than three people per square kilometer. However, 90 % of people live within 100 km of the coast (World Resources Institute: http://earthtrends.wri.org). Most people are concentrated on the east and southeast coasts (including Adelaide, Melbourne, Sydney, Brisbane, Townsville), and the southwest coast (including Perth).

Estuaries vary greatly on the Australian coast based on precipitation, evaporation, and tide regimes (Table 3.8, Figs. 3.26 and 3.27). Irregular flood and fire

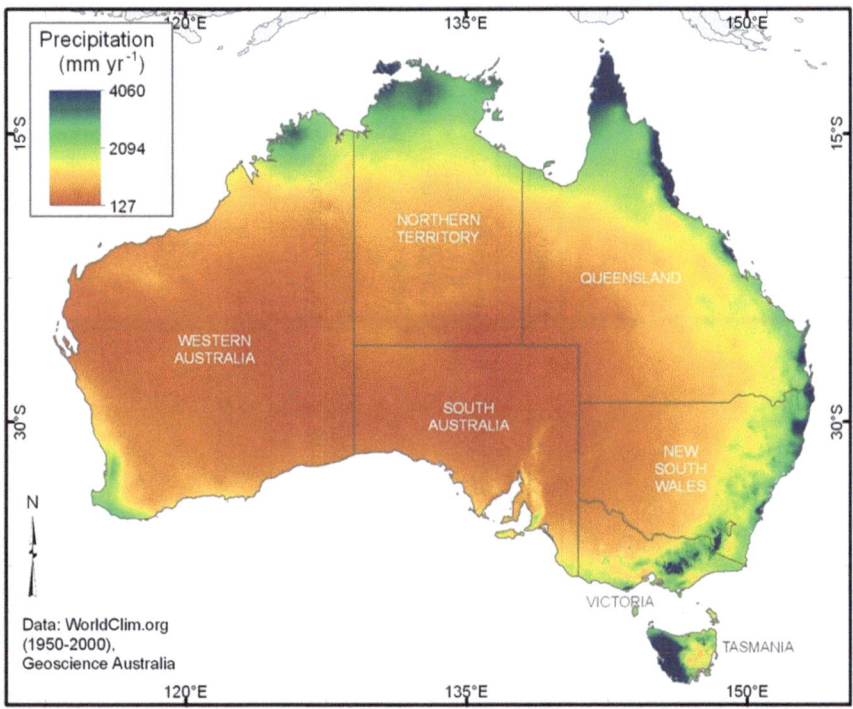

**Fig. 3.25** Mean annual rainfall in Australia from 1950 to 2000

regimes that occur in the catchments of temperate parts of Australia (mid-southern latitudes) strongly influence the water quantity and quality of the estuaries into which the catchments drain (Roy et al. 2001). Many Australian estuaries are susceptible to full or partial closure of their mouths because of the small and variable amount of precipitation in their catchments. Additional removal of freshwater for human uses can potentially harm the estuary by:

- altering the estuarine salinity gradient;
- altering estuarine flushing and water quality; and,
- obstructing diadromous fish and crustacean migration (Peirson 2002).

Many fish and crustacean species have been impacted by changes in freshwater inflows. Economically important fishes that have been affected by salinity changes include Barramundi (*Lates calcarifer*), king threadfin (*Polydactylus macrochir*) jungle perch (*Kuhlia rupestris*), Australian bass (*Macquaria novemaculeata*), estuary perch (*Macquaria colonorum*), Black Bream (*Acanthopagrus butcheri*), and blue catfish (*Arius graeffei*; Table 3.9; Cappo et al. 1998b; Robins et al. 2006; Milton et al. 2008; Halliday et al. 2008). Grazing in wetlands, sometimes in addition to barriers to fish movement, have caused water retention and increased evaporation in some Australian Estuaries, especially in Northern

**Table 3.8** Regional classification of Australian Estuaries (from Tomczak 2000)

| Location | Tidal range | Rainfall | | Estuary type | | Example |
|---|---|---|---|---|---|---|
| | | Summer | Winter | Summer | Winter | |
| East and Southeast | Moderate | Moderate | | Slightly stratified if the catchment is large enough to suppress rainfall variability; intermittent otherwise | | Hawkesbury River, Derwent River (slightly stratified), Port Hacking (intermittent) |
| South | Small | Nearly none; large evaporation | Moderate; large evaporation | Inverse | | Spencer Gulf, St. Vincent Gulf |
| Southwest | Small | A little | Moderate | Slightly stratified | Highly stratified | Swan River |
| North and Northeast | Large to extreme | Very large | Nearly none; large evaporation | Slightly stratified | Salt plug | South Alligator River, Wenlock River |

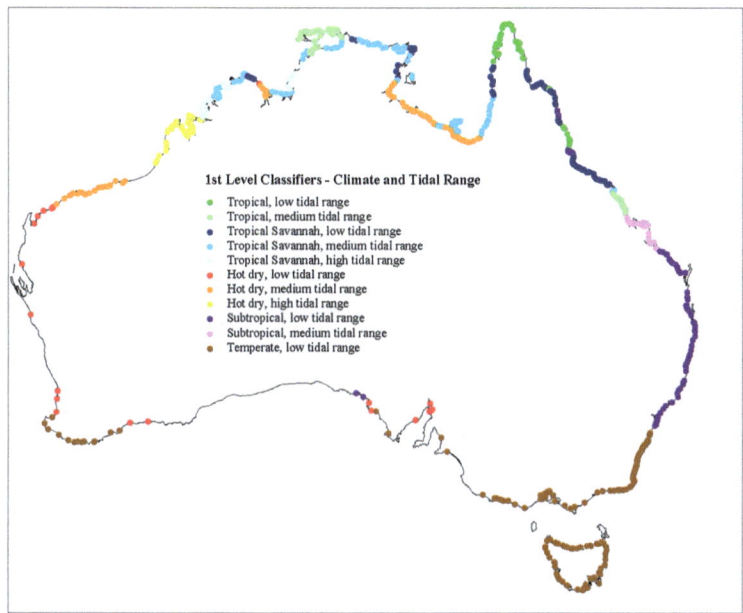

**Fig. 3.26** The geographic distribution of estuaries based on climate and tidal range (from Digby et al. 1999)

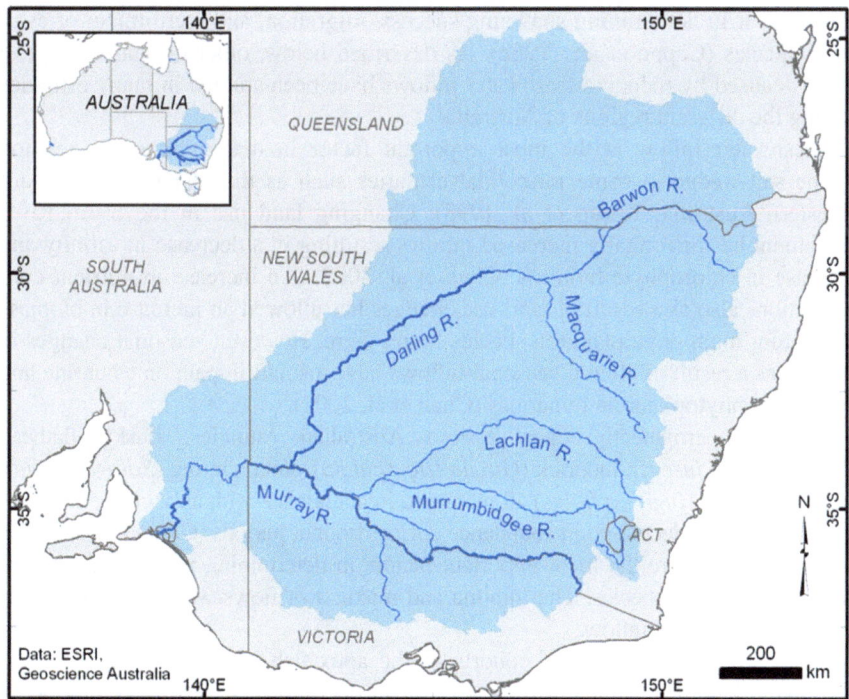

**Fig. 3.27** Murray–Darling Basin showing states and major rivers. *ACT* Australian Capital Territory

**Table 3.9** Economically important fishes of coastal freshwaters threatened by changes in drainage to estuaries (From Cappo et al. 1998b after Merrick and Schmidt 1984)

| Species | Drainage | Adult habitat | Juvenile habitat | Spawning site | Spawning salinity |
|---------|----------|---------------|------------------|---------------|-------------------|
| Barramundi | | Tidal, nontidal, longshore to headlands | Upper tidal limits and non-tidal freshwater | Sheltered estuary mouths, sandbars <2 m deep | 17–31 |
| Jungle Perch | NE Coast | Headwaters | Headwaters | Lower estuaries | Brackish |
| Australian Bass | SE Coast, NE coast (lower) | Tidal to non-tidal | Brackish water | Lower estuaries | 12–15 (larvae best at ≥ 20) |
| Estuary Perch | SE Coast, Murray Darling | Tidal, nontidal | Brackish water | Estuary mouths | |
| Blue Catfish (*Arius graeffii*) | Indian Ocean, Timor Sea, Gulf, NE Coast, SE Coast (upper) | Tidal, nontidal | Tidal, nontidal | | |

Territory (Cappo et al. 1998b). The resulting increase in salinity has caused a reduction in fish diversity, freshwater wetlands, and nutrient cycling.

The location and quality of the salinity gradient in an estuary or coastal zone is a key factor in determining spawning success, migration, and recruitment of estuarine species (Cappo et al. 1998b). As described below, other coastal ecological effects caused by reducing freshwater inflows have been studied in many estuaries among the different regions of Australia.

Freshwater inflow is the most important factor in determining the position of the salt-wedge in some microtidal estuaries such as the Swan River Estuary (Western Australia; Kurup et al. 1998). Changing land use in the Swan River catchment has historically increased inflows resulting in a decrease in salinity and increase in chlorophyte biomass (Chan et al. 2002). An increase in nutrient concentrations also as a result of land-use changes has allowed an increase in biomass of the four main phytoplankton species found there. However, seasonal changes in salinity as a result of altered seasonal inflows have a small impact on estuarine and freshwater phytoplankton dynamics (Chan et al. 2002).

In nine permanently open eastern Australian estuaries, dusky flathead (*Platycephalus fuscus*), luderick (*Girella tricuspidata*), sand whiting (*Sillago ciliata*), and sea mullet (*Mugil cephalus*) all increased in abundance with an increase in freshwater inflow and decreased in abundance during drought periods (Gillson et al. 2009). Seasonal flows were the most important factors in determining variability in abundance for the four species. The minima and maxima of flows were more important than other aspects of inflow.

Increases and decreases of cohorts of the apex fish Mulloway (*Argyrosomus japonicas*) have been related to increases and decreases in freshwater inflows in the Murray River estuary suggesting that the estuary may provide important refuge for juveniles (Ferguson et al. 2008).

In the Upper Derwent Estuary, a drowned river estuary in Tasmania, hydroelectric storage operations have decreased the frequency of flows needed to flush the already severely degraded estuary (Davies and Katish 1994). This lack of flushing has allowed low dissolved oxygen concentrations and high hydrogen sulfide concentrations in bottom waters to persist for 7 months of the year.

Different methods have been used in an attempt to determine environmental flow requirements in different regions of Australia (Pierson et al. 2002). Pierson et al. (2002) provided a multidisciplinary adaptive management methodology for assessing freshwater inflows using a checklist of major ecological processes that needs to be developed for each estuary of coastal zone. This methodology relies on the availability of high-quality ecological and water quality data, which is often lacking in estuarine systems. In tropical (northern) Australia, fisheries production has been promoted as a suitable indicator of freshwater inflows such as in the Fitzroy River Estuary in northern Queensland (Robins et al. 2005). Fisheries can be a useful indicator both because they are generally considered by the public to be important and there is generally available historical data in the form of commercial catch. The spatial and temporal extent of biological and water quality data in Victoria forced preliminary analyses of inflow requirements for the state to be based on physical measures such as GIS-based hydrological modeling and an EFLOWS hydrology modeling (Hardie et al. 2006; Barton et al. 2008). Other potential and already implemented environmental flow requirements are summarized by Gippel (2002) and Gippel et al. (2009).

### 3.7.1 The Murray–Darling River Basin

The Murray–Darling Basin (MDB) comprises the catchments of Australia's three largest rivers, the Darling River (2,740 km), the Murray River (2,530 km; also known as River Murray and Mighty Murray), the Murrumbidgee River (1,690 km), and their tributaries (Australian Bureau of Statistics 2008; Fig. 3.27). The MDB covers over one million $km^2$ (14 % of Australia's area) across five Australian states and territories (Australian Capital Territory, New South Wales, Queensland, South Australia, and Victoria). Precipitation in the MDB is spatially variable, with 1400 mm $yr^{-1}$ falling in the highlands and only 300 mm $yr^{-1}$ falling in the northwest (MDMC 1987). The high spatial variability in rainfall combined with high temporal variability that occurs, results in considerable seasonal and year-to-year variability in river flows within the MDB (Hatton MacDonald and Young 2002). Of the over half a million gigaliters of rainfall per year that falls in the MDB, 94 % evaporates, 2 % evaporates and 4 % ultimately becomes runoff. As a consequence of the high evaporation rate, 80 % of land in the MDB is classified as having an arid or semiarid climate (GWP 2003). Two-thirds of the land in the MDB is used for growing crops or is in pasture, which provides 39 % of the Australian national income that is derived from agricultural production. 2 % of the agricultural land in the MDB (1.65 million hectares) is irrigated. This equates to 65 % of all irrigated land in Australia occurring in the MDB. The 2 million people that inhabit the MDB use 52 % of Australia's total water consumption through agricultural, industrial and household uses.

The high demand for water has consequently been associated with a large built infrastructure and diversion rate, which has altered many parts of the MDB including widespread soil erosion, river siltation, accelerated recharge of groundwater aquifers and subsequent discharge of saline groundwater to rivers, increased dryland salinity, loss of flora and fauna habitat (GWP 2003), and changes to the geomorphology of the Murray River mouth (Harvey 1988, 1996; Walker and Jessup 1992; Bourman and Barnett 1995). Increases in salinities in the Murray River Estuary have been linked to changes in physiochemical processes and to changes in the abundance and distribution of flora and fauna. Changes in estuarine salinity have been linked to changes in abundance and distribution of fish species such as hardyhead species (*Craterocephalus* sp; Wedderburn et al. 2007) and mulloway (*Argyrosomus japonicas*; Ferguson et al. 2008), abundance, distribution, and diversity of plant and macroinvertebrate communities (Geddes 2004), and the spatiotemporal variation in community structure of wetland birds (Paton et al. 2009). The degradation has resulted in ecological, cultural, social, and economic consequences. Management problems in the MDB as identified by GWP (2003) include:

- increasing competition for scarce water resources
- resistance to further land clearing controls by State Governments
- increasing conflict over who should pay for remediation of degraded common resources
- how to best mobilize and target the use of available resources for on-ground action and
- how to address poorly specified institutional arrangements for common property resource management.

## 3.7.2  History of Murray–Darling Management

After a series of droughts and increasing conflict between those concerned with using the river for navigation and those concerned with using the river for irrigation, the River Murray Waters Agreement (RMWA) was signed in 1915 by the governments of Australia, New South Wales, Victoria, and South Australia. As a result of this agreement, several lakes, dams, weirs, and locks were created along the Murray River over the next 70 years by the River Murray Commission (RMC). The increasing awareness of riverine water quality (in addition to water quantity), and the connectivity between practices in the watershed and water quality on the Murray River from the 1960s to the 1980s brought about new water management responsibilities for the RMC (Table 3.10). Despite amendments to the RMWA up until the early 1980s, it became increasingly obvious that the RMWA was insufficient in managing both the dwindling resources and growing environmental degradation in the Murray River. The insufficiency of the RMWA to address especially water

**Table 3.10** Evolution of the River Murray Waters Agreement, 1914–1981 (from Clarke 1982 and http://www2.mdbc.gov.au/about/history_mdbc.html)

*Matters beyond the powers of the Commission in 1914*

Problems arising on tributary rivers

Problems caused by adjacent land use

Problems of flood mitigation and protection

Problems of erosion and catchment protection

Problems of water quality and pollution from agricultural and other sources

Problems of influent and effluent waters

The needs of flora and fauna

Possible recreational, urban, or industrial use

The environment or esthetic consequences of particular proposals

*Matters permitted by previous amendments and informal practice before 1976*

Limited powers of catchment protection

Power to initiate future proposals

Provision of certain dilution flows to maintain water quality

Lock maintenance work, improving navigability

Provision of recreational facilities

Expenditure on salinity investigations

Expenditure on re-designed works to protect fish life

Construction and operation of storages on tributaries

*Principal innovations in agreement reached in October, 1981*

Power to consider any or all relevant water management objectives including water quality, in the investigation, planning and operation of works

Power to monitor water quality

Power to coordinate studies concerning water quality in the River Murray

Power to recommend water quality standards for adoption by the states

Power to make recommendations to any government agency or tribunal on any matter which may affect the quantity or quality of River Murray waters

Power to make representations to any government agency concerning any proposal which may significantly affect the flow, use, control or quality of River Murray waters

Power to recommend future changes to the Agreement

New water accounting provisions

---

quality especially problems with salinity, can be partially attributed to its' 'limited geographical and functional scope' (Crabb 1988). To mitigate this problem, the Murray–Darling Basin Ministerial Council (MDBMC) was formed in 1985.

The MDBMC was comprised of parliamentary ministers (federal level), who were primarily responsible for the land, water, and environmental sectors of Australia, and the three relevant state governments (New South Wales, Victoria, and South Australia). In 1987, the MDBMC-directed Murray–Darling Basin Environmental Resources Study was completed (MDBMC 1987). This study was completed to determine the extent of degradation within the MDB. Also in 1987, the Murray–Darling Basin Agreement (MDBA) was signed; however, this was replaced by an entirely new agreement of the same name in 1992. The purpose of the MDBA was to "promote and co-ordinate effective planning and management for the equitable efficient and sustainable use of the water, land and other environmental resources of the Murray-Darling Basin." The new MDBA was given full legal status by the Murray–Darling Basin Act in 1993. The MDBA now involved three separate institutions to manage the MDB:

- The Murray–Darling Basin Ministerial Council (to provide policy and direction)
- The Murray–Darling Basin Committee (to manage tributaries within the catchment and advise the ministerial council) and
- The Community Advisory Group (to advise the MDBMC from a community viewpoint)

The state of Queensland and the Australian Capital Territory were included in the Agreement in 1996 and 1998, respectively.

Restructuring of the management system occurred in 2008 with the implementation of the Murray–Darling Basin Authority, which assumed the responsibilities of the former Murray–Darling Basin Committee. The Murray–Darling Basin Authority is directed by decisions made by the new six-member Authority, Ministerial Council, and the Basin Officials Committee. The key functions of the Murray–Darling Basin Authority as stated by the Authority include:

- Preparing the Basin Plan for adoption by the Minister for Climate Change, Energy Efficiency, and Water, including setting sustainable limits on water that can be taken from surface and groundwater systems across the Basin
- Advising the minister on the accreditation of state water resource plans
- Developing a water rights information service which facilitates water trading across the Murray–Darling Basin
- Measuring and monitoring water resources in the Basin
- Gathering information and undertaking research
- Engaging the community in the management of the Basin's resources (MDBA 2010).

Currently the Murray–Darling Basin Authority is developing a Basin Plan as required by the (Australian) Water Act 2007. Functions of the plan include determining and enforcing sustainable limits of surface and ground waters, set basin-wide water quality (including salinity) objectives, set requirements to be met by state resource plans, develop efficient water trading schemes, and improve water security for all uses of basin water resources (MDBA 2010). These functions will be met after considerable scientific and socioeconomic evaluations of environmental, economical, and cultural (see Morgan et al. 2004; Venn and Quiggin 2007; Steenstra 2009) basin resources.

There has been continuing speculation about the effects of the plan, particularly on agricultural practices of irrigating farmers, water allocation buy-back programs, and effects of climate change on water availability. The quantity of water that is necessary to provide satisfactory environmental conditions in the MDB is widely known to be over allocated (DEWHA 2010). This overallocation has prompted several willing-seller water rights buy-back schemes such as the Murray–Darling Basin Authority's 'Living Murray Water Purchase Project' and the Australian Government's 'Small Block Irrigators Exit Grant Package'. The Living Murray Water Purchase Project's aim to recover 500 Gl of water permits from large-scale water users (mostly irrigating farmers) involved the reviews of seller-proposed water permit prices over a 2-month period in mid-2009. The Small Block Irrigators Exit Grant Package is a plan primarily for smaller irrigating farmers (<40 ha) that was available from late 2008 until mid-2009. For further information on water buy-back in the MDB see Qureshi et al. 2010, Crase (2009) and the Department of the Environment, Water, Heritage and the Arts website (www.environment.gov.au/water).

The negative effects of the forecasted drier conditions as a result of global climate change have also been discussed (CSIRO 2008; Quiggin 2008; Adamson et al. 2009; Connor et al. 2009). Scenarios of surface water reductions in MDB include a median decline of 11 % throughout the basin, which would result in a 24 % reduction in current outflow from the Murray River mouth, and 30 % of the flow that would occur without development within the MDB (CSIRO 2008). The combined effects of climate change, surface water reductions, and overallocation provides a further stress on the already stressed MDB environment.

## 3.8 South Africa

South African law recognizes basic human water requirements as well as the need to sustain healthy freshwater and estuarine ecosystems. The Water Law Principles of 1996 set the direction of the future of water resources management (Department of Water Affairs and Forestry 2004). The twin principles of sustainability and equity run through the South Africa National Water Policy of 1997 and the National Water Act (Act 36 of 1998). The key to balancing sustainability and equity lies in the provisions for the Reserve, the ability to quantify a Reserve and the ability to manage water uses to meet the Reserve.

The concept of the Ecological Reserve is central to identifying environmental flow requirements. The policy identifies four different levels of assessment of the Ecological Reserve have been identified (Table 3.11):

- Desktop estimate (to obtain a low confidence value for the reserve of a water resource for use in the Water situation assessment model) (Not applicable to estuaries)

**Table 3.11** Potential uses for different levels of Reserve determinations

| Level | Use |
| --- | --- |
| Desktop estimate | For use in Water Situation Assessment Model (WSAM) as part of planning processes only |
| Rapid determination | Individual licensing for small impacts in unstressed catchments of low importance & sensitivity; compulsory licensing "holding action" |
| Intermediate determination | Individual licensing in relatively unstressed catchments |
| Comprehensive determination | All compulsory licensing. In individual licensing, for large impacts in any catchment. Small or large impacts in very important and/or sensitive catchments |

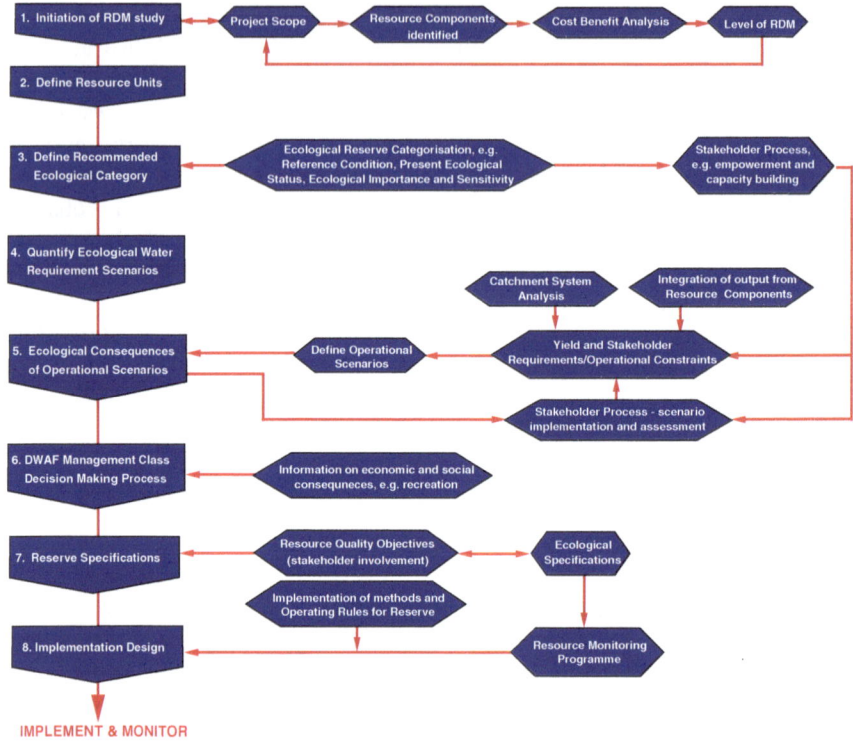

**Fig. 3.28**   Generic procedures for the determination of Resource Directed Measures

- Rapid determination
- Intermediate determination
- Comprehensive determination.

Criteria for the selection of the appropriate level of RDM determination include:

- Degree to which the catchment is already utilized
- Sensitivity and importance of a catchment, and
- Potential impact of proposed water use.

The generic procedures for determining an Ecological Reserve are shown in Fig. 3.28.

## 3.9   Mthatha Estuary

The 1998 South Africa National Water Act requires a reserve to satisfy basic human needs and to protect aquatic ecosystems. The basic human needs reserve is the right of every person to 25 l of water of adequate quality per day. In addition, the Act establishes the Ecological Reserve to protect rivers, wetlands, estuaries, and groundwater.

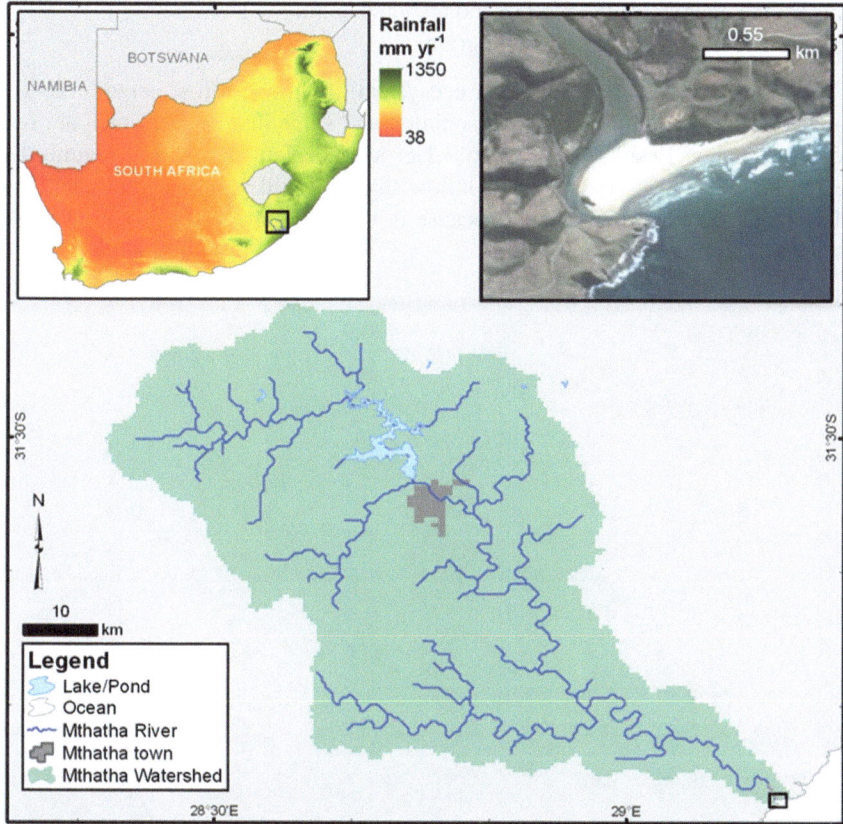

**Fig. 3.29**   Mthatha Estuary, South Africa

An example to implementing provisions of the Act is the approach used to identify inflow requirements for the Mthatha Estuary (Fig. 3.29). In the watershed, there is a storage capacity that is 50 % of mean runoff and only 8 % of the runoff now reaches the sea. The approach to determine the Ecological Reserve is an interesting example of adding explicit value judgments to the process of setting minimum flows (Adams et al. 2002). A multistep process was adopted where values (i.e., expert opinions) are of primary consideration. The method includes documenting geographic boundaries of the estuary and determining estuarine health by comparing it to a national rating system. Altogether a seven-step process was used:

1. Delineate geographic boundaries.
2. Ecoregional typing.
3. Assess present state and reference condition.
4. Determine present ecological status and importance using ecological health and importance indices.
5. Determine ecological management class

6.  Set the quantity of the reserve and resource quality objectives.
7.  Design resource monitoring program.

The more pristine, or healthy, an ecosystem, the more inflow is required. The evaluations are based on best expert opinion and value judgments, which are rendered as scores. Results are used to select an acceptable river flow regime that represents the highest reduction in inflow that will still protect the estuary. The Mthatha Estuary had high scores because it was an Ecological Reserve so it was assigned a high inflow requirement.

# Chapter 4
# Summary: Water Supply, People, and the Future

At this point in time, there is no question as to whether freshwater inflow is important to coastal ecosystems. Rather, the important questions are how, when, where, and in what quantities inflow should be used for environmental purposes (JCSCWEF 2004). Scientific methods and protocols have advanced over the past 40 years to improve our understanding of the importance of freshwater inflow (Fig. 1.2). Past work has shown that adaptive management and precautionary principle methods must be incorporated into the scientific study, management strategy implementation, and regulatory permitting phases of future freshwater inflow studies (JCSCWEF 2004; Montagna et al. 2009). This important conclusion should be applied coast-wide throughout the world to develop future approaches of managing freshwater inflow for adequate protection of coastal ecosystems.

As the famous Frenchman Victor Hugo said "science has the first word on everything, but the last word on nothing". This satiric yet true statement outlines the importance of always having a stakeholder driven, integrated ecosystem-based management process. So if the key to using the scientific framework in Fig. 1.4 is to work backwards, then stakeholder involvement in choosing the estuarine resources to be protected is a key first step (Fig. 4.1).

Again, Alber (2002) recognized that managing freshwater flows has roles for politicians, resource managers, scientists, and the public (Fig. 1.3). An important feature of Alber's conceptual model is that it is essentially an adaptive management process because managers can modify decisions based on information flow (the switch in Fig. 4.1) from scientists, policy makers, and stakeholders.

As an example, the process outlined in Fig. 4.1 is essentially the same adaptive management process that has been applied to determining the environmental flow regimes that are needed to maintain estuarine productivity for Nueces River and the Nueces Delta, Texas, USA (Montagna et al. 2009). When the Choke Canyon Reservoir was constructed on the Nueces River in 1982, a special condition required the City of Corpus Christi to provide not less than 185 million m$^3$ (151,000 ac-ft) of water per year to the Nueces Estuary through a combination of spills, releases, and return flows to maintain ecological health and productivity of

P. A. Montagna et al., *Hydrological Changes and Estuarine Dynamics*,
SpringerBriefs in Environmental Science, DOI: 10.1007/978-1-4614-5833-3_4,
© The Author(s) 2013

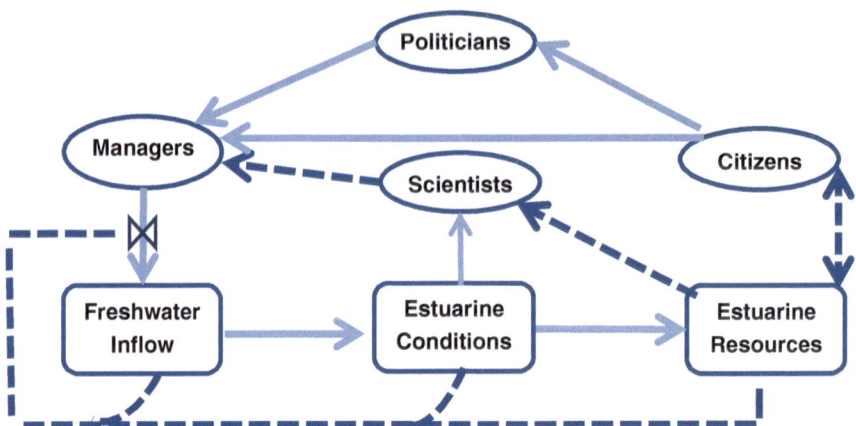

**Fig. 4.1** Conceptual model of inflow management with roles of stakeholders. *Solid lines* denote control and *dashed lines* denote information flow. *Switch* denotes adaptive management

living marine resources. However, no releases were made and salinities in Nueces Bay increased to hypersaline conditions during the drought period of 1988–1990. The estuary was inverted where fresher seawater entered the delta and diluted the hypersaline parts of the estuary only on high tides. In contrast, the bay histori- cally supported populations of shrimp and oysters generally requiring salinities in the range of 10–20 psu. Since 1990, a series of restoration activities, monitoring and experimental studies, and other amendments to the water permit has resulted in a stakeholder driven process that increased environmental health while provid- ing a sustainable water supply to the region. The lesson learned is that the goal of achieving consensus can occur with a blend and balance of science and a stake- holder decision process (Fig. 4.1).

The relationship between inflow and estuarine resources is surely a problem of biocomplexity. Further adding to the complexity are the long period climate cycles (e.g., the El Niño-Southern Oscillation, the Pacific Decadal Oscillation, and the North Atlantic Multi-decadal Oscillation), which cause lags in biological responses to current events (Estevez 2002; Pollack et al. 2011). After reviewing many inflow studies, Estevez (2002) came to three conclusions: (1) flow-effects studies will be improved if they occur at multiple trophic levels, multiple scales, and multiple temporal periods, (2) within-estuary salinity gradients have to be exploited because they control structure and function, but care must be taken because of nonlinear relationships, and (3) it must be determined if instream analyses can inform inflow recommendations. The third conclusion is perhaps the most understudied and potentially important issue. Essentially, it asks: To what extent are riverine flows protective of estuarine resources? This is important because it directly points to the conjunctive use and exploitation of rivers and estuaries. Is it possible that if we "take care of rivers" then "we have already taken care of estuaries?"

In this book, we have summarized the current state of knowledge as it relates to altered hydrology of rivers and how that alters estuarine function. Undoubtedly, we have missed a few key studies. However, the case studies proffered here demonstrate a few key concepts. We now understand that unlike rivers, there are no direct influences of inflow in estuaries. Instead, there is a domino effect where flow drives water quality conditions, and water quality drives estuarine living resources. It is impossible to manage freshwater inflow without a policy framework and an adaptive management process.

Once there is a mandate for creating inflow needs to estuaries, there are many different methodologies to choose from. The important considerations in choosing an appropriate methodological approach include: (1) whether data are available to apply the methodology in a scientifically sound manner; (2) whether results of applying the methodology are reasonably accurate; and (3) whether the results can be translated into flow recommendations that can be implemented by the responsible parties. By conducting studies that meet these three considerations, scientists will be able to more accurately determine the ecological and socioeconomical impacts of changing freshwater inflows. In turn, stakeholders and managers will then be able to make well-informed decisions to successfully manage freshwater inflows to their local coasts.

# References

Adams JB, Bate GC, Harrison TD, Huizinga P, Taljaard S, Van Niekerk L, Plumstead EE, Whitfield AK, Wooldridge TH (2002) A method to assess the freshwater inflow requirements of estuaries and application to the Mtata Estuary, South Africa. Estuaries 25:1382–1393

Adamson D, Mallawaarachchi DT, Quiggin J (2009) Declining inflows and more frequent droughts in the Murray–Darling Basin: climate change, impacts and adaptation. Aust J Agric Res Econ 53:345–366

Alber M (2002) A conceptual model of estuarine freshwater inflow management. Estuaries 25:1246–1261

Allan EL, Froneman PW, Hodgson AN (2006) Effects of temperature and salinity on the standard metabolic rate (SMR) of the caridean shrimp *Palaemon peringueyi*. J Expt Mar Biol Ecol 337:103–108

Alveirinho JMA, Gonzalez R, Ferreira O (2004) Natural versus anthropic causes in variations of sand export from river basins: an example from the Guadiana river mouth (Southwestern Iberia). Polish Geol Inst Spec Pap 11:95–102

Andree S (ed) (1983) Apalachicola oyster industry: conference proceedings. Florida Sea Grant College, 85 pp

Aragón-Noriega EA, Calderón-Aguilera LE (2000) Does damming of the Colorado River affect the nursery area of blue shrimp *Litopenaeus stylirostris* (Decapoda: Penaeidae) in the Upper Gulf of California? Revista Biol Trop 48:867–871

Atrill MJ, Rundle SD, Thomas RM (1996) The influence of drought-induced low freshwater flow on an upper-estuarine macroinvertebrate community. Water Res 30:261–268

Australian Bureau of Statistics (2008) 4610.0.55.007—Water and the Murray–Darling Basin—A Statistical Profile, 2000-01 to 2005-06. www.abs.gov.au. Accessed 15 Aug 2008

Banse K, Mosher S (1980) Adult body mass and annual production/biomass relationships of field populations. Ecol Monogr 50:355–379

Barton JL, Pope AJ, Quinn GP, Sherwood JE (2008) Identifying threats to the ecological condition of Victorian estuaries. Department of Sustainability and Environment Technical Report. Victoria, Australia

Bates, BC, Kundzewicz ZW, Wu S, Palutikof JP (eds) (2008) Climate change and water. Technical Paper of the Intergovernmental Panel on Climate Change, IPCC Secretariat, Geneva, 210 pp

Bennett WA (2005) Critical assessment of the delta smelt population in the San Francisco Estuary, California. San Franci Estuary Watershed Sci [online serial] 3(2), Article 1. http://www.escholarship.org/uc/item/0725n5vk?display=all. Accessed 12 Feb 2010

Bourman RP, Barnett EJ (1995) Impacts of river regulation on the terminal lakes and mouth of the River Murray, South Australia. Aust Geogr Stud 33:101–115

Bricker SB, Clement CG, Pirhalla DE, Orlando SP, Farrow DRG (1999) National estuarine eutrophication assessment: effects of nutrient enrichment in the nation's estuaries. NOAA,

P. A. Montagna et al., *Hydrological Changes and Estuarine Dynamics*,
SpringerBriefs in Environmental Science, DOI: 10.1007/978-1-4614-5833-3,
© The Author(s) 2013

National Ocean Service, Special Projects Office and the National Centers for Coastal Ocean Science, Silver Spring, Maryland

Brock DA (2001) Nitrogen budget for low and high freshwater inflows, Nueces Estuary, Texas. Estuaries 24:509–521

Browder JA, Moore D (1981) A new approach to determining the quantitative relationship between fishery production and the flow of fresh water to estuaries. In: Cross RD, Williams DL (eds) Proceedings of the national symposium on freshwater inflow to estuaries, vol 1. U.S. Fish and Wildlife Service, U.S. Department of Interior, Washington, D.C.

Browder JA, Zein-Eldin Z, Criales MM, Robblee MB, Wong S, Jackson TL, Johnson D (2002) Dynamics of pink shrimp (*Farfantepenaeus duorarum*) recruitment potential in relation to salinity and temperature in Florida Bay. Estuaries 25:1355–1371

Bunn S, Arthington A (2002) Basic principles and ecological consequences of altered flow regimes for aquatic biodiversity. Environ Manag 30:492–507

Bureau of Reclamation (2000) Concluding Report: Rincon Bayou Demonstration Project. Volume I: Executive Summary. United States Department of the Interior, Bureau of Reclamation, Oklahoma-Texas Area Office, Austin, Texas

CALFED Bay-Delta Program (2010) http://calwater.ca.gov/index.aspx. Accessed 23 March 2012

California's Colorado River Water Use Plan (2000) Colorado River Board of California. The Resources Agency, State of California. http://www.crb.ca.gov/Calif_Plan%May%20 11%Draft.pdf. Accessed 23 March 2012

Cappo M, Alongi DM, Williams DM, Duke N (1998a) A review and synthesis of Australian fisheries habitat research: Major threats, issues and gaps in knowledge of coastal and marine fisheries habitats—Volume 1: A prospectus of opportunities for the FRDC "ecosystem protection program" FRDC 95/055. Australian Institute of Marine Science

Cappo M, Alongi DM, Williams DM, Duke N (1998b) A review and synthesis of Australian fisheries habitat research: Major threats, issues and gaps in knowledge of coastal and marine fisheries habitats—Volume 2: Scoping Review FRDC 95/055. Australian Institute of Marine Science

Carter NT, Corn ML, Abel A, Kaplan SM, Buck EH, Brougher C, Alexander K (2008) Apalachicola-Chattahoochee-Flint (ACF) drought: federal water management issues. CRS report for congress. Congressional Research Service. Order code RL34326

Central Coast Region Basin Plan (2006) Central Coast Regional Water Quality Control Board

Central Valley Basin Plan (2009) Central Valley Regional Water Quality Control Board. http://www.waterboards.ca.gov/centralvalley/water_issues/basin_plans/. Accessed 23 March 2012

CH2M Hill (1999) Water supply needs and sources assessment: alternative water supply investigation: Review of established minimum flows and levels for the Wekiva River system. Special Publication SJ99-SP1. St. John's River Water Management District, Palatka, Florida

Chadwick MA, Feminella JW (2001) Influence of salinity and temperature on the growth and production of a freshwater mayfly in the Lower Mobile River, Alabama. Limnol Oceanogr 46:532–542

Chan TU, Hamilton DP, Robson BJ, Hodges BR, Dallimore CJ (2002) Impacts of hydrological changes on phytoplankton succession in the Swan River, Western Australia. Estuaries 25:1405–1415

Chícharo MA, Chícharo L, Galvão H, Barbosa A, Marques MH, Andrade JP, Esteves E, Miguel C, Gouveia C (2001) Status of the Guadiana estuary (South Portugal) during 1996–1998: an ecohydrological approach. Aquat Ecosyst Health Manag 4:1–17

Chícharo L, Chícharo MA, Esteves E, Andrade P, Morais P (2002) Effects of alterations in fresh water supply on the abundance and distribution of *Engraulis encrasicolus* in the Guadiana Estuary and adjacent coastal areas of south Portugal. J Ecohydrol Hydrobiol 1:195–200

Chícharo L, Chícharo MA, Ben-Hamadou R (2006a) Use of a hydrotechnical infrastructure (Alqueva dam) to regulate planktonic assemblages in the Guadiana estuary: basis for the sustainable water and ecosystem services management. Estuar Coast Shelf Sci 70:3–18

Chícharo MA, Chícharo L, Morais P (2006b) Interannual differences of ichthyofauna structure of the Guadiana estuary and adjacent coastal area (SE Portugal/SW Spain): before and after Alqueva dam construction. Estuar Coast Shelf Sci 70:39–51

Chícharo MA, Leitão T, Range P, Gutierrez C, Morales J, Morais P, Chícharo L (2009) Alien species in the Guadiana Estuary (SE-Portugal/SW-Spain): *Blackfordia virginica* (Cnidaria, Hydrozoa) and *Palaemon macrodactylus* (Crustacea, Decapoda): potential impacts and mitigation measures. Aquat Invasions 4:501–506

Chollett I, Bone D (2007) Effects of heavy rainfall on polychaetes: differential spatial patterns generated by a large-scale disturbance. J Exp Mar Biol Ecol 340:113–125

Cisneros-Mata MA, Montemayor-Lopez G, Roman-Rodriguez MJ (1995) Life history and conservation of *Totoaba macdonaldi*. Conserv Biol 9:806–814

Clark SD (1982) The River Murray Waters agreement: down the drain or up the creek? Civil Eng Trans 24:201–208

Colorado River Basin Plan (2006) Colorado River Basin Regional Water Quality Control Board. http://www.waterboards.ca.gov/coloradoriver/publications_forms/publications/docs/basinplan _2006.pdf. Accessed 3 March 2012

Colorado River Compact (1922) US Department of the Interior, Bureau of Reclamation. http://www.usbr.gov/lc/region/g1000/pdfiles/crcompct.pdf. Accessed 3 March 2012

Connor J, Schwabe K, King D, Kaczan D, Kirby M (2009) Impacts of climate change on lower Murray irrigation. Aust J Agric Res Econ 53:437–456

Copeland BJ (1966) Effects of decreased river flow on estuarine ecology. J Water Pollut Control Fed 38:1831–1839

Costanza R, d'Arge R, de Groots R, Farber S, Grasso, Hannon B, Limburg K, Naeem S, O'Neill RV, Paruelo J, Raskins RG, Sutton P, van den Belt M (1997) The value of the world's ecosystem services and natural capital. Nature 387:253–260

Cowie G (ed) (2002) Reservoirs in Georgia: meeting water supply needs while minimizing impacts. River Basin Science and Policy Center, University of Georgia. 34 pp

Crabb P (1988) Managing the Murray–Darling Basin. Aust Geogr 19:64–88

Crase L, O'Keefe S, Dollery B (2009) Water Buy-Back in Australia: political, technical and allocative challenges. A Contributed Paper to the Australian Agricultural & Resource Economics Society's Annual Conference, Cairns, 11–13 February 2009

Cross R, Williams D (eds) (1981) Proceedings of the national symposium on freshwater inflow to estuaries. U.S. Fish and Wildlife Service, Office of Biological Services. FWS/OBS-81/04, 2 vols

Crossland CJ, Kremer HH, Lindeboom HJ, Marshall Crossland JI, Le Tissier MDA (eds) (2005) Coastal fluxes in the anthropocene. Springer, Berlin

CSIRO (2008) Water availability in the Murray–Darling Basin. Summary of a report to the Australian Government from the CSIRO Murray-Darling Basin Sustainable Yields Project. CSIRO, Australia, 12 pp. www.csiro/mdbsy. Accessed 13 March 2010

Culliton T.J. (1998) Population: Distribution, Density and Growth. NOAA's State of the Coast Report. Silver Spring, MD. NOAA. http://state-of-coast.noaa.gov:80/bulletins/html/pop_01/ references.html

Dahms HU (1990) Salinity, temperature, pH tolerance and grain-size preference of *Paramphiascella fulvofasciata* (Copepoda, Harpacticoida). Biol Jb Dodonaea 58:48–62

Davies JL (1973) Geographical variation in coastal development. Hafner, New York

Davies PE, Katish SR (1994) Influence of river hydrology on the dynamics and water quality of the Upper Dewent Estuary, Tasmania. Aust J Mar Freshw Res 45:109–130

Day JW, Hall CAS, Kemp WM, Yáñez-Arancibia A (1989) Estuarine ecology. Wiley, New York

Department of Water Affairs and Forestry (2004) Water resource protection and assessment policy implementation process. Resource directed measures for protection of water resource: Methodology for the Determination of the Ecological Water Requirements for Estuaries. Version 2, Final Draft. Pretoria

DEIS (1998) Draft Environmental Impact Statement. Water Allocation for the Apalachicola-Chattahoochee-Flint River Basin: Alabama, Florida and Georgia. US Army Corps of Engineers, Mobile District. http://www.sam.usace.army.mil/pd/actacfeis/. Accessed 3 March 2010

Delta Science Program (2010) Delta Stewardship Council. http://www.deltacouncil.ca.gov/delta_ science_program/. Accessed 4 March 2012

DEWHA (Department of the Environment, Water, Heritage and the Arts) (2010) Murray Darling Basin. www.environment.gov.au/water/locations/murray-darling-basin. Accessed 28 Feb 2010

Dias JMA, Gonzalez R, Ferreira Ó (2004) Natural versus anthropic causes in variations of sand export from river basins: an example from the Guadiana River mouth (Southwestern Iberia). Polish Geol Instit Spec Pap 11:95–102

Diener RA (1975) Cooperative Gulf of Mexico Estuarine Inventory and Study-Texas: Area Description. U.S. Department of Commerce, NOAA Tech. Rep. NMFS Circ. 393, 125 pp

Digby MJ Saenger P, Whelan MB. McConchie D. Eyre B. Holmes N, Bucher D (1999) A Physical Classification of Australian Estuaries. LWRRDC Occasional Paper 16/99 (National River Health Program, Urban Sub–Program, Report No. 9), Land and Water Resources Research and Development Corporation, Canberra, ACT, Australia. au.riversinfo.org/library/nrhp/estuary_clasifn/. Accessed 3 March 2010

Doering PH, Chamberlin RH, Haunert DE (2002) Using submerged aquatic vegetation to establish minimum and maximum freshwater inflows to the Caloosahatchee Estuary, Florida. Estuaries 25:1343–1354

Domingues RB, Sobrino C, Galvão H (2007) Impact of reservoir filling on phytoplankton succession and cyanobacteria blooms in a temperate estuary. Est Coast Shelf Sci 74:31–43

Drake P, Arias AM, Baldo F, Cuesta JA, Rodriguez A, Silva-Garcia A, Sobrino I, Garcia-Gonzalez D, Fernandez-Delgado C (2002) Spatial and temporal variation of the nekton and hyperbenthos from a temperate European estuary with regulated freshwater inflow. Estuaries 25:451–468

Drinkwater KF, Frank KT (1994) Effects of river regulation and diversion on marine fish and invertebrates. Aquat Cons Mar Freshw Ecosyst 4:135–151

Dynesius M, Nilsson C (1994) Fragmentation and flow regulation of river systems in the northern third of the world. Science 266:753–762

ECT (2008) Environmental evaluations for the development of minimum flows and levels for the St. John's River near Deland at State Road 44, Volusia County. Special Publication SJ2004-SP30. St. John's River Water Management District, Palatka, Florida

Erzini K (2005) Trends in NE Atlantic landings (southern Portugal): identifying the relative importance of fisheries and environmental variables. Fish Oceanogr 14:195–209

Estevez ED (2002) Review and assessment of biotic variables and analytical methods used in estuarine inflow studies. Estuaries 25:1291–1303

Estevez ED, Marshall MJ (1993) Sebastian River salinity regime. Mote Marine Laboratory Special Publication SJ94-SP1. Mote Marine Laboratory, Sarasota, Florida

Faria A, Morais P, Chícharo MA (2006) Ichthyoplankton dynamics in the Guadiana estuary and adjacent coastal area, South-East Portugal. Estuar Coast Shelf Sci 70:85–97

Ferguson GJ, Ward TM, Geddes MC (2008) Do historical catches and recent age structures of mulloway, *Argyrosomus japonicus*, reflect freshwater inflows in the remnant estuary of the Murray River, South Australia? Aquat Living Res 21:145–152

Ferraris JD, Fauchald K, Kensley B (1994) Physiological responses to fluctuation in temperature or salinity in invertebrates—adaptations of *Alpheus viridari* (Decapoda, Crustacea), *Terebellides parva* (Polychaeta) and *Golfingia cylindrata* (Sipunculida) to the mangrove habitat. Mar Biol 120:397–406

Feyrer F, Nobriga ML, Sommer TR (2007) Multidecadal trends for three declining fish species: habitat patterns and mechanisms in the San Francisco Estuary, California, USA. Can J Fish Aquat Sci 64:723–734

Finney CM (1979) Salinity stress in harpacticoid copepods. Estuaries 2:132–135

FL DEP (2010) Florida Department of Environmental Protection. Apalachicola-Chattahoochee-Flint River System Timeline of Action as of July 27, 2009. http://www.dep.state.fl.us/mainpage/acf/timeline.htm. Accessed 24 Feb 2010

Flannery MS, Peebles EB, Montgomery RT (2002) A percent-of-flow approach for managing reductions of freshwater inflows from unimpounded rivers to southwest Florida estuaries. Estuaries 25:1318–1332

Florida Administrative Code (2005) Section 62.40-473. Minimum Flows and Levels. https://www.flrules.org/gateway/ruleNo.asp?ID=62-40.473. Accessed 3 March 2010

Fradkin PL (1981) A river no more: the colorado river and the west. University of California Press, Berkeley

Geddes M (2004) Survey to investigate the ecological health of the North and South Lagoons of the Coorong, June/July 2003. Report submitted to Upper South East Environmental Management Advisory Group. South Australian Research and Development Institute, 21 pp

Gillson J, Scandol J, Suthers I (2009) Estuarine gillnet fishery catch rates decline during drought in eastern Australia. Fish Res 99:26–37

Gippel C (2002) Workshop on environmental water requirements for Australian estuaries. Outcomes and further directions. Report to Environment Australia

Gippel CJ, Anderson B, Harty C, Bond N, Sherwood J, Pope A (2009) Gap analysis and strategy development for national level estuary environmental flows policies. Waterlines report, National Water Commission, Canberra, Australia

Glenn E, Garcia J, Tanner R, Congdon C, Luecke D (1999) Status of wetlands supported by agricultural drainage water in the Colorado River Delta, Mexico. Hortic Sci 34:16–21

Glenn EP, Zamora-Arroyo F, Nagler PL, Briggs M, Shaw W, Flessa K (2001) Ecology and conservation biology of the Colorado River Delta, Mexico. J Arid Environ 49:5–15

Glenn E, Lee C, Fegler R, Zengel S (1996) Effects of water management on the wetlands of the Colorado River Delta, Mexico. Conserv Biol 10:1175–1186

GWP (Global Water Partnership) (2003) 'The Murray-Darling Basin Commission, Australia (Case # 25)'. Integrated Water Resources Management Toolbox. www.gwptoolbox.org. Accessed 3 March 2010

Halliday IA, Robins JB, Mayer DG, Staunton-Smith J, Sellin MJ (2008) Effects of freshwater flow on the year-class strength of a non-diadromous estuarine finfish, king threadfin (*Polydactylus macrochir*), in a dry-tropical estuary. Mar Freshw Res 59:157–164

Hardie R, Lloyd L, Sherwood J (2006) Determining the environmental water requirements of Victoria's estuaries—development of a module to extend the FLOWS methodology—draft method for pilot trials. Earth Tech, Melbourne, Australia

Harvey N (1988) Coastal management issues for the mouth of the River Murray, South Australia. Coast Manag 16:139–149

Harvey N (1996) The significance of coastal processes for management of the River Murray estuary. Aust Geogr Stud 34:45–57

Hatton MacDonald D, Young MD (2001) A case study of the Murray–Darling Basin: Final report for the International Water Management Institute. Policy and Economic Research Unit, CSIRO Land and Water, Adelaide, Australia

Hess M (2010) Freshwater inflow management issues in Texas: What's being done?. Presentations at Freshwater Inflows 2010 and Beyond Conference, Corpus Christi, February 2010. http://www.freshwaterinflows2010.org/FinalProceedings.pdf. Accessed 10 June 2012

Hoese HD (1967) Effects of higher than normal salinities on salt marshes. Contrib Mar Sci 12:249–261

Istanbul Water Guide (2009) Outcomes of the 5th World Water Forum. Istanbul. htt p://www.worldwatercouncil.org/fileadmin/wwc/World_Water_Forum/WWF5/ global_water_framework_part_1_final.pdf

IBWC (2000) Minute 306: Conceptual framework for United States—Mexico studies for future recommendations concerning the riparian and estuarine ecology of the Limitrophe section of the Colorado River and its associated delta. Amends United States-Mexico, 59 Stat. 1219. Treaty with Mexico Respecting the Utilization of the Waters of the Colorado and Tijuana Rivers and of the Rio Grande, Feb 3, 1944

JCSCWEF (Joint Committee on the Study Commission on Water for Environmental Flows) (2004) Interim Report to the 79th Legislature. Available on line at http://www.twdb.state.tx.us /EnvironmentalFlows/pdfs/StudyCommFinalReport.pdf

Kalke RD, Montagna PA (1991) The effect of freshwater inflow on macrobenthos in the Lavaca River Delta and Upper Lavaca Bay, Texas. Contrib Mar Sci 32:49–71

Katz D (2006) Going with the flow: preserving and restoring instream water allocations. In: Gleik PH (ed) The World's Water 2006-2007. Island Press, Washington, pp 29–49

Keiser RK, Aldrich DV (1973) Gradient apparatus for study of salinity preference of small benthic and free swimming organisms. Contrib Mar Sci 17:153–162

Kim H-C, Montagna PA (2009) Implications of Colorado River freshwater inflow to benthic ecosystem dynamics: a modeling study. Est Coast Shelf Sci 83:491–504

Kimmerer WJ (2002) Physical, biological, and management responses to variable freshwater inflow into San Francisco Estuary. Estuaries 25:1275–1290

Kjerfve B (1979) Measurements and analysis of water current, temperature, salinity and density. In: Dyer KR (ed) Hydrography and sedimentation in estuaries. Cambridge University Press, United Kingdom, pp 186–216

Kowalewski M, Serrano GEA, Flessa KW, Goodfriend GA (2000) Dead delta's former productivity: two trillion shells at the mouth of the Colorado River. Geology 28:1059–1062

Kurup GR, Hamilton DP, Patterson JC (1998) Modelling the effect of seasonal flow variations on the position of salt wedge in a microtidal estuary. Estuar Coast Shelf Sci 47:191–208

Lahontan Region Basin Plan (1995) Lahontan Regional Water Quality Control Board. http://www.waterboards.ca.gov/lahontan/water_issues/programs/basin_plan/references.shtml

Lane NF, Bierbaum RM, Anderson MT (2003) Science and water policy for the United States. In: Lawford RG, Fort DD, Hartmann HC, Eden S (eds) Water: science, policy, and management. American Geophysical Union, Water Resource Monograph 16, Washington, pp 207–222

La Peyre MK, Gossman B, La Peyre JF (2009) Defining optimal freshwater flow for oyster production: effects of freshet rate and magnitude of change and duration on eastern oysters and Perkinsus marinus Infection. Estuar Coasts 32:522–534

Larkin TJ, Bomar GW (1983) Climatic atlas of Texas. Texas Department of Water Resources. Austin, 151 pp

Lercari D, Defeo O, Celentano E (2002) Consequences of a freshwater canal discharge on the benthic community and its habitat on an exposed sandy beach. Mar Pollut Bull 44:1397–1404

Lin C-I (2010) Apalachicola-Chattahoochee-Flint Tri-State Negotiation. Aquapedia. Tufts University. https://wikis.uit.tufts.edu/confluence/display/aquapedia/Apalachicola-Chattahoochee-Flint+Tri-State+Negotiation

Livingston RJ, Niu XF, Lewis FG, Woodsum GC (1997) Freshwater input to a Gulf estuary: long-term control of trophic organization. Ecol Appl 7:277–299

Longley WL (ed) (1994) Freshwater inflows to Texas bays and estuaries: ecological relationships and methods for determination of needs. Texas Water Development Board and Texas Parks and Wildlife Department, Austin, 386 p

Los Angeles Region Water Quality Control Basin Plan (1995) Los Angeles Regional Water Quality Control Board. http://www.waterboards.ca.gov/losangeles/water_issues/programs/basin_plan/basin_plan_documentation.shtml

Luecke D (2000) An environmental perspective on large ecosystem restoration processes and the role of the market, litigation, and regulation. Arizona Law Rev 42:395–410

Luecke DF, Pitt J, Congdon C, Glenn E, Valdés-Casillas C, Briggs M (1999) A delta once more: Washington, D.C., Environmental Defense Fund Publications. http://www.edf.org/documents/425_delta.pdf

MacKenzie CL Jr (1977) Development of an aquaculture program for rehabilitation of damaged oyster reefs in Mississippi. US Fish Wildl Serv Mar Fish Rev 39:1–13

Mackin J (1956) Dermocystidium marinum and salinity. Proc Natl Shellfish Assoc 46:116–128

Mannino A, Montagna PA (1997) Small-scale spatial variation of macrobenthic community structure. Estuaries 20:159–173

Mattson RA (2002) A resource-based framework for establishing freshwater inflow requirements for the Suwannee River estuary. Estuaries 25:1333–1342

McEachron LW, Fuls B (1996) Trends in relative abundance and size of selected finfishes and shellfishes along the Texas coast. Coastal Fisheries Division, Management Data Series, Austin, Texas

McLeod RJ, Wing TR (2008) Influence of an altered salinity regime on the population structure of two infaunal bivalve species. Est Coast Shelf Sci 78:529–540

Meade RH, Yuzyk T, Day T (1990) Movement and storage of sediment in rivers of the United States and Canada. In: Riggs WH (ed) The geology of North America. Geological Society of America

Menzel RW, Hulings NC, Hathaway RR (1958) Causes of depletion of oysters in St. Vincent Bay, Apalachicola Bay, Florida. Proc Nat Shellfish Assoc 48:66–71

Merrick JR, Schmida GE (1984) Australian freshwater fishes: biology and management. Griffin Press, Sydney

Millennium Ecosystem Assessment (MEA) (2005) Ecosystems and human well being: current state and trends. Findings of the condition and trends working group. Island Press, Washington

Milton D, Halliday I, Sellin M, Marsh R, Staunton-Smith J, Woodhead J (2008) The effect of habitat and environmental history on otolith chemistry of barramundi *Lates calcarifer* in estuarine populations of a regulated tropical river. Estuar Coast Shelf Sci 78:301–315

Montagna PA (2008) Long-term response of benthic organisms to freshwater inflow in Texas coastal bend estuaries. Final Report to Texas Parks and Wildlife Department. http://harteresea rchinstitute.org/montagna/publications.html

Montagna PA, Alber M, Doering P, Connor MS (2002a) Freshwater inflow: science, policy, management. Estuaries 25:1243–1245

Montagna PA, Kalke RD, Ritter C (2002b) Effect of restored freshwater inflow on macrofauna and meiofauna in upper Rincon Bayou, Texas, USA. Estuaries 25:1436–1447

Montagna PA, Estevez ED, Palmer TA, Flannery MS (2008a) Meta-analysis of the relationship between salinity and molluscs in tidal river estuaries of southwest Florida, USA. Am Malacol Bull 24:101–115

Montagna PA, Palmer TA, Pollack JB (2008b) St. John's Estuary: estuarine benthic macroinvertebrates. Phase 1 final report. Submitted to St. John's River Water Management District

Montagna PA, Hill EM, Moulton B (2009) Role of science-based and adaptive management in allocating environmental flows to the Nueces Estuary, Texas, USA. In: Brebbia CA, Tiezzi E (eds) Ecosystems and sustainable development VII. WIT Press, Southampton, pp 559–570

Montagna PA, Kalke RD (1992) The effect of freshwater inflow on meiofaunal and macrofaunal populations in the Guadalupe and Nueces Estuaries, Texas. Estuaries 15:307–326

Montagna PA, Kalke RD (1995) Ecology of infaunal Mollusca in south Texas estuaries. Am Malacol Bull 11:163–175

Kim H-C, Montagna PM (2012) Effects of climate-driven freshwater inflow variability on macrobenthic secondary production in Texas lagoonal estuaries: a modeling study. Ecol Model 235–236:67–80

Montagna PA, Li J (2010) Effect of freshwater inflow on nutrient loading and macrobenthos secondary production in Texas lagoons. In: Kennish MJ, Paerl HW(eds) Coastal lagoons: critical habitats of environmental change. CRC Press, Taylor & Francis Group, Boca Raton, pp 513–539

Montagna PA, Li J, Street GT (1996) A conceptual ecosystem model of the Corpus Christi Bay National Estuary Program Study Area. Publication CCBNEP-08, Texas Natural Resource Conservation Commission, Austin, Texas. http://cbbep.org/publications/virtuallibrary/ccbnep 08.pdf. Accessed 10 May 2004

Montagna PA, Ward G, Vaughan B (2011) The importance and problem of freshwater inflows to Texas estuaries. In: Griffin RC (ed) Water policy in Texas: responding to the rise of scarcity. The RFF Press, Washington, pp 107–127

Montague CL, Ley JA (1993) A possible effect of salinity fluctuation on abundance of benthic vegetation and associated fauna in northeastern Florida Bay. Estuaries 16:703–717

Morais P, Chícharo MA, Chícharo L (2009) Changes in a temperate estuary during the filling of the biggest European dam. Sci Total Environ 407:2245–2259

Morgan M, Strelein L, Weir J (2004) Indigenous rights to water in the Murray Darling Basin. AIATSIS Research Discussion Paper, vol 14. AIATSIS, Canberra

MBBMC (Murray-Darling Basin Ministerial Council) (1987) Murray-Darling Basin Environmental Resources Study. Sydney, NSW, Australia. 426 pp

MDBA (Murray-Darling Basin Authority) (2010) www.mdba.gov.au. Accessed March 2010

Naiman RJ, Magnuson JJ, McKnight DM, Stanford JA, Karr JR (1995) Freshwater ecosystems and their management: a national initiative. Science 270:584–585

National Water Policy of India (2002) Government of India, Ministry of Water Resources. New Delhi. http://wrmin.nic.in/writereaddata/linkimages/nwp20025617515534.pdf

Nichols FH, Cloern JE, Luoma SN, Peterson DH (1986) The modification of an estuary. Science 231:567–573

Normant M, Lamprecht I (2006) Does scope for growth change as a result of salinity stress in the amphipod *Gammarus oceanicus*? J Expert Mar Biol Ecol 334:158–163

North Coast Region Water Quality Control Basin Plan (2007) North Coast Regional Water Quality Control Board. http://www.waterboards.ca.gov/northcoast/water_issues/programs/basin_plan/083105-bp/02_introduction.pdf

NWFWMD (2010) Northwest Florida Water Management District. http://www.nwfwmd.state.fl.us/

OPPAGA (Office of Program Policy Analysis and Government Accountability) (1995) Performance review of the consumptive use permitting program administered by the Department of Environmental Protection and the Water Management Districts. Report No. 94-34. http://www.oppaga.state.fl.us/Summary.aspx?reportNum=94-34

Olsen SB, Padma TV, Richter BD (2007) Managing freshwater inflows to estuaries: a methods guide. U.S. Agency for International Development, Washington, 44 pp

Orlando SP Jr, Rozas LP, Ward GH, Klein CJ (1993) Salinity characteristics of Gulf of Mexico estuaries. National Oceanic and Atmospheric Administration, Office of Ocean Resources Conservation and Assessment, Silver Spring, 209 pp

Paton DC, Rogers DJ, Hill BM, Bailey CP, Ziembicki M (2009) Temporal changes to spatially stratified waterbird communities of the Coorong, South Australia: implications for the management of heterogenous wetlands. Anim Conserv 12:408–417

Pearson TH, Rosenberg R (1978) Macrobenthic succession in relation to organic enrichment and pollution in the marine environment. Oceanogr Mar Biol Ann Rev 16:229–311

Pierson WL, Bishop K, Van Senden D, Horton PR, Adamantidis CA (2002) Environmental water requirements to maintain estuarine processes. Environmental flows initiative technical report number 3, Commonwealth of Australia, Canberra

Pitt J (2001) Can we restore the Colorado River Delta? J Arid Environ 49:211–220

Pitt J, Luecke D, Cohen M, Glenn E, Valdes-Casillas C (2000) Two countries, one river: managing ecosystem conservation in the Colorado River delta. Nat Resour J 40:819–864

Poff NL, Allan JD, Bain MB, Karr JR, Prestegaard KL, Richter BD, Sparks RE, Stromberg JC (1997) The natural flow regime; a paradigm for river conservation and restoration. BioScience 47:769–784

Poff NL, Allan JD, Palmer MA, Hart DD, Richter BD, Arthington AH, Rogers K, Meyers JL, Stanford JA (2003) Fiver flows and water wars: emerging science for environmental decision making. Front Ecol Environ 1:298–306

Pollack JB, Kinsey JW, Montagna PA (2009) Freshwater inflow biotic index (FIBI) for the Lavaca-Colorado Estuary, Texas. Environ Bioindic 4:153–169

Pollack J, Palmer TA, Montagna PA (2011) Long-term trends in the response of benthic macrofauna to climate variability in the Lavaca-Colorado Estuary, Texas. Mar Ecol Prog Ser 436:67–80

Postel SL, Daily GC, Ehrlich PR (1996) Human appropriation of renewable fresh water. Science 271:785–788

Posetel S, Richter B (2003) Rivers for life. Island Press, Washington

Powell GL, Matsumoto J, Brock DA (2002) Methods for determining minimum freshwater inflow needs of Texas bays and estuaries. Estuaries 25:1262–1274

Pringle CM, Freeman MC, Freeman BJ (2000) Regional effects of hydrologic alterations on riverine macrobiota in the New World: tropical-temperate comparisons. BioScience 50:807–823

Pritchard DW (1952) Estuarine hydrography. In: Advance in geophysics, vol I. Academic Press, New York, pp 243–280

Pritchard DW (1967) What is an estuary: physical viewpoint. In: Lauff GH (ed) Estuaries. American Association for the Advancement of Science, Washington, pp 52–63

Rabalais NN, Nixon SW (2002) Preface: nutrient over-enrichment of the coastal zone. Estuaries 25:639

Republic of South Africa (1998) National Water Act, No 36 of 1998. Government Gazette, Pretoria

Rhoads DC, McCall PL, Yingst JY (1978) Disturbance and production on the estuarine seafloor. Am Sci 66:577–586

Richter BR, Mathews R, Harrison DL, Wigington R (2003) Ecologically sustainable water management: managing river flows for ecological integrity. Ecol Appl 13:206–224

Riera P, Montagna PA, Kalke RD, Richard P (2000) Utilization of estuarine organic matter during growth and migration by juvenile brown shrimp *Penaeus aztecus* in a South Texas estuary. Mar Ecol Prog Ser 199:205–216

Robins JB, Halliday IA, Staunton-Smith J, Mayer DG, Sellin MJ (2005) Freshwater-flow requirements of estuarine fisheries in tropical Australia: a review of the state of knowledge and the application of a suggested approach. Mar Freshw Res 56:343–360

Robins J, Mayer D, Staunton-Smith J, Halliday I, Sawynok B, Sellin M (2006) Variable growth rates of the tropical estuarine fish barramundi *Lates calcarifer* (Bloch) under different freshwater flow conditions. J Fish Biol 69:379–391

Robinson L, Campbell P, Butler L (2000) Trends in Texas commercial fishery landings, 1972-1998. Management Data Series No. 173. Texas Parks and Wildlife Department, Austin

Rocha C, Galvão H, Barbosa A (2002) Role of transient silicon limitation in the development of cyanobacteria blooms in the Guadiana Estuary, south-western Iberia. Mar Ecol Prog Ser 228:35–45

Rozas LP, Minello TJ, Munuera-Fernandez I, Fry B, Wissel B (2005) Macrofaunal distributions and habitat change following winter-spring releases of freshwater into the Breton Sound estuary, Louisiana (USA). Est Coast Shelf Sci 65:319–336

Ruhl JB, Lant C, Loftus T, Kraft S, Adams J, Duram L (2003) Proposal for a model state Watershed Management Act. Environ Law 33:929–947

Rutger SM, Wing SR (2006) Effects of freshwater input on shallow-water infaunal communities in Doubtful Sound, New Zealand. Mar Ecol Prog Ser 314:35–47

Sabine and Neches Rivers and Sabine Lake Bay Basin and Bay Expert Science Team (2009) Environmental Flows Recommendations Report: Final Submission to the Sabine and Neches Rivers and Sabine Lake Bay Basin and Bay Area Stakeholder Committee, Environmental Flows Advisory Group, and Texas Commission on Environmental Quality. Submitted November 2009

San Diego Basin Plan (1994) San Diego Regional Water Quality Control Board. http://www.waterboards.ca.gov/sandiego/water_issues/programs/basin_plan/index.shtml

San Francisco Basin (Region 2) Water quality control plan (Basin plan) (2007) California Regional Water Quality Control Board, San Francisco Bay Region. http://www.swrcb.ca.gov/sanfranciscobay/water_issues/programs/basin_plan/docs/bp_ch1withcover.pdf

Santa Ana Region Basin Plan (2008) Santa Ana Regional Water Quality Control Board. http://www.waterboards.ca.gov/santaana/water_issues/programs/basin_plan/index.shtml

Saoud IP, Davis DA (2003) Salinity tolerance of brown shrimp *Farfantepenaeus aztecus* as it relates to postlarval and juvenile survival, distribution, and growth in estuaries. Estuaries 26:970–974

Science Advisory Committee (2004) Report On Water For Environmental Flows. Prepared for Senate Bill 1639, 78th Legislature, Study Commission On Water For Environmental Flows, October 26, 2004. http://www.tceq.state.tx.us/assets/public/permitting/watersupply/water_rights/txefsac8132008article4.pdf. Accessed 31 March 2010

Science Advisory Committee (2006) Recommendations of the Science Advisory Committee. Presented to Governor's Environmental Flows Advisory Committee, August 21, 2006. http://www.tceq.state.tx.us/assets/public/permitting/watersupply/water_rights/txefsac8132008article5.pdf. Accessed 31 March 2010

Science Advisory Committee (2009a) Use of hydrologic data in the development of instream flow recommendations for the environmental flows allocation process and the Hydrology-Based Environmental Flow Regime (HEFR) Methodology. Report # SAC-2009-01-Rev1., April 20, 2009. http://www.tceq.state.tx.us/assets/public/permitting/watersupply/water_rights/eflows/hydrologicmethods04202009.pdf. Accessed 31 March 2010

Science Advisory Committee (2009b) Geographic scope of instream flow recommendations. Report # SAC-2009-02, April 3, 2009. http://www.tceq.state.tx.us/assets/public/permitting/watersupply/water_rights/eflows/geographicscope.pdf. Accessed 31 March 2010

Science Advisory Committee (2009c) Methodologies for establishing a freshwater inflow regime for Texas estuaries within the context of the Senate Bill 3 environmental flows process. Report # SAC-2009-03-Rev1., June 5, 2009. http://www.tceq.state.tx.us/assets/public/permitting/watersupply/water_rights/eflows/fwi20090605.pdf. Accessed 31 March 2010

Science Advisory Committee (2009d) Fluvial sediment transport as an overlay to instream flow recommendations for the environmental flows allocation process. Report # SAC-2009-04, May 29, 2009 http://www.tceq.state.tx.us/assets/public/permitting/watersupply/water_rights/eflows/sac_2009_04_sedtransport.pdf. Accessed 31 March 2010

Science Advisory Committee (2009e) Essential steps for biological overlays in developing Senate Bill 3 instream flow recommendations. Report # SAC-2009-05, August 31, 2009. http://www.tceq.state.tx.us/assets/public/permitting/watersupply/water_rights/eflows/biologyoverlay.pdf. Accessed 31 March 2010

Science Advisory Committee (2009f) Nutrient and water quality overlay for hydrology-based instream flow recommendations. Report # SAC-2009-06, November 3, 2009. http://www.tceq.state.tx.us/assets/public/permitting/watersupply/water_rights/eflows/wqoverlay_200911039.pdf. Accessed 31 March 2010

Science Advisory Committee (2010a) Discussion Paper: Moving from instream flow regime matrix development to environmental flow standard recommendations. February 17, 2010. http://www.tceq.state.tx.us/assets/public/permitting/watersupply/water_rights/eflows/sac_discussionpaper.pdf. Accessed 31 March 2010

Science Advisory Committee (2010b) Lessons learned from initial SB3 BBEST Activities. Report # SAC-2010-01, July 14, 2010. http://www.tceq.state.tx.us/assets/public/permitting/watersupply/water_rights/eflows/20100714sac_lessonslearned.pdf. Accessed 31 March 2010

Science Advisory Committee (2010c) Considerations in the development of an SB3 work plan for adaptive management. Report # SAC-2010-02, August 20, 2010. http://www.tceq.state.tx.us/assets/public/permitting/watersupply/water_rights/eflows/20100820sac_guidance_workplan.pdf. Accessed 31 March 2010

Schmitt RJ, Osenberg CW (1996) Detecting ecological impacts: concepts and applications in coastal habitats. Academic Press, San Diego

SFWMD (South Florida Water Management District) (2010) http://www.sfwmd.gov/

SJRWMD (St. John's River Water Management District) (2005) District Water Management Plan, September 2005. http://sjr.state.fl.us/dwmp/index.html

SJRWMD (St. John's River Water Management District) (2010) Available online and accessed 31 March 2010 at http://sjr.state.fl.us/

SJRWMD (2012) The St. Johns River Water Supply Impact Study. http://www.sjrwmd.com/surfacewaterwithdrawals/impacts.html. Accessed 4 Aug 2012

Sommer T, Armor C, Baxter R, Breuer R, Brown L, Chotkowski M, Culberson S, Feyrer F, Gingras M, Herbold B, Kimmerer W, Mueller-Solger A, Nobriga M, Souza K (2007) The collapse of pelagic fishes in the Upper San Francisco Estuary. Fisheries 32:270–277

Soule DF (1988) Marine organisms as indicators: reality or wishful thinking? In: Soule DF, Kleppel GS (eds) Marine organisms as indicators. Springer, New York, pp 1–11

SRWMD (Suwannee River Water Management District) (2010) http://www.srwmd.state.fl.us/

State Water Resources Control Board (2010) California Environmental Protection Agency. http://www.waterboards.ca.gov/about_us/water_boards_structure/whoweare.shtml

Steenstra A (2009) Accommodating Indigenous Cultural Values in Water Resource Management: The Waikato River, New Zealand; the Murray- Darling Basin, Australia; and the Colorado

River, USA. Australian Agricultural & Resource Economics Society's Annual Conference, Cairns, 11–13 Feb 2009

Stevens LE, Ayers TJ, Bennett JB, Christensen K, Kearsley MJC, Meretsky VJ, Phillips AM, Parnell RA, Spence J, Sogge MK, Springer AE, Wegner DL (2001) Planned flooding and Colorado River riparian trade-offs downstream from Glen Canyon Dam, Arizona. Ecol Appl 11:701–710

Stevens LE, Schmidt JC, Brown BT (1995) Flow regulation, geomorphology, and Colorado River marsh development in the Grand Canyon, Arizona. Ecol Appl 5:1025–1039

Swarzenski PW, Reich CD, Spechler RM, Kindinger JL, Moore WS (2001) Using multiple geochemical tracers to characterize the hydrogeology of the submarine spring off Crescent Beach, Florida. Chem Geol 179:187–202

SWFWMD (Southwest Florida Water Management District) (2005) District Water Management Plan, July 2005. http://www.swfwmd.state.fl.us/about/watermanagementplan/

SWFWMD (Southwest Florida Water Management District) (2010) http://www.swfwmd.state.fl.us/

Syvitski JPM, Vörösmarty CJ, Kettner AJ, Green P (2005) Impact of humans on the flux of terrestrial sediment to the global coastal ocean. Science 308:376–380

Texas Department of Water Resources (1982) The influence of freshwater inflows upon the major bays and estuaries of the Texas Gulf coast. Texas Department of Water Resources, report LP-115, Austin, Texas. 53 p, plus appendices

Texas Parks and Wildlife (1988) Trends in Texas commercial fishery landings, 1977-1987. Management Data Series, No. 149. Texas Parks and Wildlife Department, Coastal Fisheries Branch, Austin, Texas

Thompson H (2006) Water law. A practical approach to resource management and the provision of services. Juta & Co Ltd, Cape Town, 769 pp

Thompson RW (1968) Tidal flat sedimentation on the Colorado River Delta, Northwestern Gulf of California. Memoir 107. Geological Society of America

Tolan JM (2007) El Niño-Southern Oscillation impacts translated to the watershed scale: estuarine salinity patterns along the Texas Gulf coast, 1982 to 2004. Estuar Coast Shelf Sci 72:247–260

Tolley SG, Volety AK, Savarese M, Walls LD, Linardich C, Everham EM (2006) Impacts of salinity and freshwater inflow on oyster reef communities in Southwest Florida. Aquat Living Resour 19:371–387

Treaty Series 994 (1944) Utilization of waters of the Colorado and Tijuana Rivers and of the Rio Grande. US Department of the Interior, Bureau of Reclamation. http://www.usbr.gov/lc/region/g1000/pdfiles/mextrety.pdf

Trinity and San Jacinto Rivers and Galveston Bay Basin and Bay Expert Science Team (2009) Environmental flows recommendations report: Final submission to the Trinity and San Jacinto Rivers and Galveston Bay Basin and Bay Area Stakeholder Committee, Environmental Flows Advisory Group, and Texas Commission on Environmental Quality. Submitted November 2009

USACE (United States Army Corp of Engineers) (1998) draft environmental impact statement. Water allocation for the Apalachicola-Chattahoochee-Flint River Basin: Alabama, Florida and Georgia. US Army Corps of Engineers, Mobile District. http://www.sam.usace.army.mil/pd/actacfeis/

U.S. Census Bureau (2010) http://www.census.gov/

Valdez RA, Hoffnagle TL, McIvor CC, McKinney T, Leibfried WC (2001) Effects of a test flood on fishes of the Colorado River in Grand Canyon, Arizona. Ecol Appl 11:686–700

Valiela I (1995) Marine ecological processes, 2nd edn. Springer, New York

Varady RG, Hankins KB, Kaus A, Young E, Merideth R (2001) … to the Sea of Cortes: nature, water, culture, and livelihood in the Lower Colorado River basin and delta - an overview of issues, policies, and approaches to environmental restoration. J Arid Environ 49:195–209

Venn TJ, Quiggin J (2007) Accommodating indigenous cultural heritage values in resource assessment: Cape York Peninsula and the Murray-Darling Basin, Australia. Ecol Econ 61:334–344

Vörösmarty CJ, Meybeck M, Fekete B, Sharma K, Green P, Syvitski JPM (2003) Anthropogenic sediment retention: major global impact from registered river impoundments. Glob Planet Change 39:169–190

Vörösmarty CJ, Sahagian D (2000) Anthropogenic disturbance of the terrestrial water cycle. BioScience 50:753–765

Walker DJ, Jessup A (1992) Analysis of the dynamic aspects of the River Murray mouth, South Australia. J Coast Res 8:71–76

Ward GH, Irlbeck MJ, Montagna PA (2002) Experimental river diversion for marsh enhancement. Estuaries 25:1416–1425

Water Framework Directive (2000) Directive 2000/60/EC of the European parliament and of the council establishing a framework for Community action in the field of water policy. http://eur-lex.europa.eu/LexUriServ/LexUriServ.do?uri=CONSLEG:2000L0060:20011216:EN:PDF

Wedderburn SD, Walker KF, Zampatti BP (2007) Habitat separation of Craterocephalus (Atherinidae) species and populations in off-channel areas of the lower River Murray, Australia. Ecol Freshw Fish 16:442–449

Wilber DH (1992) Associations between freshwater inflows and oyster productivity in Apalachicola Bay, Florida. Estuar Coast Shelf Sci 35:179–190

Woodhouse CA, Gray ST, Meko DM (2006) Updated streamflow reconstructions for the Upper Colorado River Basin. Water Resour Res 42:W05415. doi:10.1029/2005WR004455

Zein-Eldin ZP (1963) Effect of salinity on growth of postlarval penaeid shrimp. Biol Bull 125:188–196

Zein-Eldin ZP, Aldrich DV (1965) Growth and survival of postlarval Penaeus aztecus under controlled conditions of temperature and salinity. Biol Bull 129:199–216

Zeng W, Jiang F, Zhang Y (2009) Reservoir management in the Apalachicola-Chattahoochee-Flint (ACF) river system during the interim operation plan (IOP) during the ongoing drought. In: Proceedings of the 2009 Georgia water resources conference, April 27–29, University of Georgia